Nanotechnology: Delivering on the Promise Volume 1

ACS SYMPOSIUM SERIES **1220**

Nanotechnology: Delivering on the Promise Volume 1

H. N. Cheng, Editor
U.S. Department of Agriculture, Agricultural Research Service
New Orleans, Louisiana

Laurence Doemeny, Editor
San Diego, California

Charles L. Geraci, Editor
National Institute for Occupational Safety and Health
Cincinnati, Ohio

Diane Grob Schmidt, Editor
2015 President, American Chemical Society
Washington, DC

American Chemical Society, Washington, DC

Distributed in print by Oxford University Press

Library of Congress Cataloging-in-Publication Data

Names: Cheng, H. N., editor.
Title: Nanotechnology : delivering on the promise Volume 1 / H.N. Cheng, U.S. Department of Agriculture, Agricultural Research Service, New Orleans, Louisiana [and three others], editor.
Description: Washington, DC : American Chemical Society, [2016] | Series: ACS symposium series ; 1220 | Includes bibliographical references and index.
Identifiers: LCCN 2016029727 (print) | LCCN 2016037275 (ebook) | ISBN 9780841231313 | ISBN 9780841231306 (ebook)
Subjects: LCSH: Nanotechnology.
Classification: LCC T174.7 .N37364 2016 (print) | LCC T174.7 (ebook) | DDC 620/.5--dc23
LC record available at https://lccn.loc.gov/2016029727

The paper used in this publication meets the minimum requirements of American National Standard for Information Sciences—Permanence of Paper for Printed Library Materials, ANSI Z39.48n1984.

Distributed in print by Oxford University Press

Foreword

The ACS Symposium Series was first published in 1974 to provide a mechanism for publishing symposia quickly in book form. The purpose of the series is to publish timely, comprehensive books developed from the ACS sponsored symposia based on current scientific research. Occasionally, books are developed from symposia sponsored by other organizations when the topic is of keen interest to the chemistry audience.

Before agreeing to publish a book, the proposed table of contents is reviewed for appropriate and comprehensive coverage and for interest to the audience. Some papers may be excluded to better focus the book; others may be added to provide comprehensiveness. When appropriate, overview or introductory chapters are added. Drafts of chapters are peer-reviewed prior to final acceptance or rejection, and manuscripts are prepared in camera-ready format.

As a rule, only original research papers and original review papers are included in the volumes. Verbatim reproductions of previous published papers are not accepted.

ACS Books Department

Contents

Indexes

Foreword

Nanotechnology represents a huge R&D investment worldwide. The global funding of nanotechnologies was estimated to be about $7 billion in 2011 and has increased about 20% per year since then, according to various studies. The U.S. is certainly investing heavily in nanotechnology. It started the National Nanotechnology Initiative (NNI) about 16 years ago, pulling together the efforts of 20 federal departments and independent agencies. Since 2001, the NNI has spent a total of $22 billion; in fiscal 2015 the NNI had a budget of $1.5 billion. Many exciting programs have been initiated by the NNI, and several commercial products have appeared in the market.

As a scientist myself, I am very interested in nanotechnology and the promise that it holds. Thus, when I was elected to the American Chemical Society (ACS) Presidential Succession, I made nanotechnology part of my scientific platform during my Presidential year in 2015. A major action item was to organize a high-profile Presidential symposium on nanotechnology to be held at the ACS national meeting in Denver in 2015. It was gratifying that the symposium was well attended with leading researchers and representatives of U.S. funding and regulatory agencies sharing valuable information and updates. My thanks to all the speakers and especially Larry Doemeny and Chuck Geraci who played leading roles in helping me organize this successful symposium.

The second part of my nanotechnology initiative was to initiate an ACS book on nanotechnology. The purpose of the two volumes of this book is to compile the latest R&D findings and to assess the current progress towards commercialization of nanotech products. I am pleased that H. N. Cheng, Larry Doemeny and Chuck Geraci were willing to collaborate with me to co-edit these books. We have invited the speakers from the Denver symposium and also many other well-known nanotechnology experts to contribute articles to these books. I am delighted that this book project has come to fruition. Thanks are due to all of the authors and my co-editors for their wonderful efforts.

The two volumes of this book contain a wealth of information on research, product development, commercialization, and regulatory issues related to nanotechnology. My hope is that these books will be a valuable resource and reference for students and active practitioners alike.

Diane Grob Schmidt
2015 ACS President

Preface

The two volumes of this book are based on the successful Presidential symposium on "Nanotechnology: Delivering on the Promise", which took place at the ACS Spring national meeting in Denver in March 2015. The symposium featured leading researchers in U.S. academia, industry and government labs as well as representatives of major U.S. funding and regulatory agencies. The topics covered included cutting-edge research on nanotechnology and useful information on product development, manufacturing and commercialization of selected nanotech products. Their reports noted impressive progress in all aspects of nanotechnology with a huge (and growing) number of publications and numerous products already in the marketplace. However, there were also some cautionary voices on the health, safety and environmental issues related to nanotechnology. Concurrently, all of the symposium speakers were invited to provide chapters to this ACS book. We were pleased that many of them did. Additional nanotechnology experts were also invited to submit manuscripts in order to strengthen the coverage of specific technical areas.

A total of 25 chapters are included in the two volumes of the book. Volume 1 contains two sections. The first section (Managing Nanotechnology R&D) consists of chapters written by well-known leaders at academic institutions as well as several leaders at U.S. government agencies, namely, the National Nanotechnology Coordination Office, the National Science Foundation, the Department of Defense, and the National Institutes of Health. They provide wonderful insight on how nanotechnology R&D is being managed and some of the results of their efforts.

In the second section (Health, Safety, Environmental, and Regulatory Issues), several leading experts share their perspectives and experiences with health, safety, environmental, and regulatory issues relating to nanotech products. In product development today, it is a priority to address these concerns. Furthermore, a good knowledge of the trends in government policy and regulatory issues is a distinct competitive advantage in business and product development.

In the second volume of this book, many notable applications of nanotechnology are delineated; these include new developments in energy and electronics, materials, bio/medical areas, and agriculture.

We appreciate the efforts of all the authors who took time to prepare their manuscripts and the many reviewers for their time and talent during the peer review process. Thanks are also due to the personnel at ACS Books, particularly Arlene Furman, Elizabeth Hernandez, Bob Hauserman, and Mary Calvert for their efficient handling of the manuscripts. We hope these volumes will stand out as a source of up-to-date information for researchers, a useful guide for practitioners interested in product development, and a handy reference for people who want to learn more about nanotechnology.

H. N. Cheng, PhD
Southern Regional Research Center
USDA Agricultural Research Service
1100 Robert E. Lee Blvd.
New Orleans, LA 70124
hn.cheng@ars.usda.gov

Laurence J. Doemeny, PhD
San Diego, CA
ldoemeny@cox.net

Charles L. Geraci, PhD
National Institute for Occupational Safety & Health (NIOSH)
1090 Tusculum Avenue, C-14
Cincinnati, OH 45226
ciu9@cdc.gov

Diane Grob Schmidt, PhD
2015 ACS President
American Chemical Society
1155 Sixteenth Street, N.W.
Washington, DC 20036
diane.g.schmidt@uc.edu

Chapter 1

Nanotechnology Overview: Opportunities and Challenges

H. N. Cheng,[1] Laurence J. Doemeny,[2] Charles L. Geraci,[3] and Diane Grob Schmidt[*,4]

[1]USDA Agricultural Research Service, Southern Regional Research Center, 1100 Robert E. Lee Boulevard, New Orleans, Louisiana 70124, United States
[2]4922 Armin Way, San Diego, California 92115, United States
[3]National Institute for Occupational Safety and Health (NIOSH), 1090 Tusculum Avenue, C-14, Cincinnati, Ohio 45226, United States
[4]Department of Chemistry, University of Cincinnati, 301 Clifton Court, Cincinnati, Ohio 45221, United States
[*]E-mail: diane.g.schmidt@uc.edu.

Nanotechnology can be defined as the science of manipulating matter at the nanometer scale in order to discover new properties and possibly produce new products. For the past 30 years, a considerable amount of scientific interest and R&D funding devoted to nanotechnology has led to rapid developments in all areas of science and engineering, including chemistry, materials, energy, medicine, biotechnology, agriculture, food, electronic devices, and consumer products. In the U.S. alone, the federal government has spent more than $22 billion in nanotechnology research since 2001. Already some products have appeared in the marketplace and more will certainly come in the future. A possible concern is the health, safety, and environmental impact of some of these products.

Many of the opportunities and challenges of nanotechnology are described in the following chapters. This chapter serves as an overview and aims to summarize the key information reported in those chapters.

Introduction

The Organisation for Economic Co-operation and Development (OECD) defines nanotechnology as a "set of technologies that enables the manipulation, study or exploitation of very small (typically less than 100 nanometers) structures and systems" (*1*). This field is broad and applies to many disciplines of research, including chemistry, materials, energy, medicine, biotechnology, agriculture, food, electronics, magnetic, optics, and information technology. A large number of books and review articles are available on this topic; several of them dealing with product development and commercialization are given in the References (*2–11*). Many of the opportunities and challenges provided by nanotechnology are described in the chapters given in both volumes of this book (*12–35*).

The key feature of nanotechnology is the size of the materials involved. Typically, a carbon nanotube may have a diameter of about 2 nm, quantum dots (e.g., colloidal semiconductor nanocrystals) can be as small as 2-10 nm, and nanocellulose can vary in size from 1 nm to hundreds of nanometers. For comparison, the diameter of an atom is 0.1 to 0.5 nm, a protein may have a size that varies from 1 to 10 nm, DNA has a size about 2 nm, a virus has a diameter of about 20-400 nm, whereas a bacterial cell may be about 1000 nm in diameter. The small sizes of nanomaterials have opened up new opportunities for product developments, prompted many new research and development (R&D) projects, and spawned numerous publications worldwide.

In the U.S., President Bill Clinton announced the creation of the National Nanotechnology Initiative (NNI) in 2000 involving 20 federal departments and independent agencies working together toward the shared vision of "a future in which the ability to understand and control matter at the nanoscale leads to a revolution in technology and industry that benefits society" (*13*, *36*). With the support of the NNI, nanotechnology R&D is thriving in academic, government, and industry laboratories across the U.S. The NNI has spent $22 billion since 2001 and has a fiscal 2015 budget of $1.5 billion. The $1.5 billion is broken down into multiple component areas (*37*):

- 34% in foundational research
- 25% in application devices
- 16% in infrastructure and instrumentation
- 7% in environmental health and safety
- 18% in signature initiatives – areas the government is trying to focus on, with specific goals over a multi-year period, e.g., solar energy, nanomanufacturing, nanoelectronics, knowledge infrastructure, and sensors

A large number of successful R&D programs and commercial ventures have started. Interested readers may want to check out the following document: "Report to the President and Congress on the Fifth Assessment of the National Nanotechnology Initiative", available on the web (*38*).

R&D activities in nanotechnology outside of the U.S. are also high. The global funding of nanotechnologies was estimated to be about $7 billion in 2011 (*39*, *40*) and has increased at about 20% per year since then. The countries that invested heavily in nanotechnology include China, Japan, Germany, Russia, France, South Korea, and Great Britain. There is no question that nanotechnology is a hot field today.

Planning and Managing Nanotechnology R&D

With so many resources devoted to nanotechnology R&D, appropriate planning and management is necessary for success. In this book, the critical roles assumed in the U.S. by the NNI, National Science Foundation (NSF), Department of Defense (DOD), and National Institutes of Health (NIH) are reported by Fadel and Meador (*13*), Roco (*14*), Sloter (*15*) and Henderson (*16*), respectively.

According to Roco (*14*), progress in nanotechnology may be divided into three stages: Nano 1 (~2000-2010) focused on uncovering phenomena at the nanoscale and semi-empirical synthesis of nanocomponents; Nano 2 (~2010-2020) on science-based nanosystem integration for fundamentally new products; and Nano 3 (~2020-2030) on creating new nanosystem architectures and converging technology platforms. Already global revenues of nano-enabled products have reached $1 trillion in 2013 according to industry surveys, and annual U.S. and global revenue growth rate has increased from about 25% during 2001-2010 to about 40% during 2010-2014 (*15*).

Since 2001, over $22 billion have been invested in fundamental and applied nanotechnology R&D through the NNI. In the chapter by Fadel and Meador (*13*), the structure and history of the NNI is highlighted with a few examples of advancement in nanoscience, and a synopsis is provided of what the future may hold for the initiative and nanoscience in the U.S. The NNI is now transitioning towards a critical inflection point with greater focus placed on the rapid commercialization of nanotechnologies.

Sloter (*15*) in his chapter presents the status and ongoing role of nanomaterials and manufacturing research and evolving applications in the context of the DOD's national security mission. He provides highlights of materials and processes associated with the military departments and defense agencies, particularly enhanced or new defense capabilities and advances in engineered nanomaterials. DOD is also a founding participant of the NNI.

Henderson (*16*) reports that NIH has invested about $4.2 billion since 2001 in nanotechnology research. In addition to its participation in the NNI, NIH has also supported extramural research with respect to specially formulated initiatives or as investigator-initiated grants and contracts. For example, three bioengineering initiatives were released in 2002 to promote the development and application of nanotech tools to solve biomedical problems. In 2005, three new nanotech initiatives with significant set-aside funds were launched, involving nanomedicine, diagnostics and therapeutics for cancer, and diagnosis and treatment of heart, lung, blood, and sleep disorders. NIH also participates in the federal-wide Precision Medicine Initiative launched in 2015.

In addition to the federal government, many U.S. universities are also investing heavily in nanotech. There are many ways whereby universities can organize their resources and infrastructures in order to maximize the outcome. Two approaches are discussed in the chapter by Liehr (*17*). The first is to create spin-off companies based on the intellectual property developed by the faculty and students. At the State University of New York Polytechnic Institute (SUNY Poly), the SUNY Research Foundation facilitates this process and administers externally sponsored research and technology transfer. The second approach is to foster public-private partnerships, with close collaboration between universities, the state government, and companies located at or close to the campuses, typically via industrial parks. At SUNY Poly, the initial focus is on nano-electronics, and it has been successful in installing state-of-the-art facilities and developing numerous joint programs with industrial partners.

Mirkin describes a third approach (*12*). In 2000, Northwestern University set up the International Institute for Nanotechnology (IIN) to be an umbrella organization that unifies endeavors in nanotechnology by researchers at different parts of Northwestern and neighboring Chicago-area research institutions, including Argonne National Laboratory. They focus on several areas, such as medicine, materials and devices, energy, and information technology. So far, their work has led to a combined R&D throughput of over $800 million, created over 100 partnerships in 18 countries, established 36 different research centers, and seen the participation of over 750 graduate students and postdocs and over 190 faculty members. A partnership with Northwestern's Kellogg School of Management has facilitated the start-up of 20 different companies and attracted over $700 million in venture capital.

Promising Outcomes and Developments

The potential opportunities of nanotechnology in product development are fully illustrated in several chapters in this book. For convenience, these are grouped into the following categories: 1) Energy and Electronics, 2) Materials, 3) Bio/Medical, 4) Food and Agriculture. Brief descriptions are given in the following sections.

Opportunities in Energy and Electronics

Lewis (*18*) provides an excellent review of a viable artificial photosynthetic system. It consists of two complementary, current-matched and voltage-adding photosystems, in conjunction with two different catalysts: one to oxidize water, and the other to reduce either water and/or carbon dioxide to solar fuels. The light-absorbing semiconductors are designed and grown as high-aspect-ratio microwires which simultaneously allow minimization of ionic transport pathways, sufficient depth for light absorption in the semiconductor, efficient collection of charge carriers, and high surface areas for catalyst loading. Non-noble-metal catalysts for the redox reactions have been discovered and methods developed for protecting the semiconductors against corrosion. This progress towards a robust, efficient, inexpensive and safe solar-fuels generator provides a good example of nanoscale materials-by-design.

In their chapter, Rajeeva and Zheng (*19*) point out the promise and the utility of nanophotonics. In metal nanostructures, collective oscillations of conductive electrons lead to nanoscale confinement of optical fields with high thermal energy. The interactions of these effects with molecules (known as molecular plasmonics) have stimulated the design of novel devices for a variety of applications. Their work in three areas is reviewed: molecular-scale measurements and control, directed and self-assembled nanofabrication, and real-life applications. They further summarize two unconventional nanofabrication techniques: moiré nanosphere lithography and multi-photon plasmonic lithography, which can be utilized for real-life applications.

Iocozzia and Lin (*20*) have designed novel "nanoreactors" that can generate well-defined organic nanostructures. They have produced nanoscale devices for applications in photovoltaics, energy storage, photocatalysis, ferroelectricity and optics. Each device requires a combination of techniques, including templating, nanoscale interfacial modification, placement, and tuning. Whereas nanocomposite design is far more complicated than that for bulk nanocomposites, they have achieved promising properties with their devices.

Opportunities in Materials

In their chapter, Qi et al. (*21*) review their approach to combine polymer single crystals and nanoparticles to produce new types of hybrid materials. The functionalized single crystals are of interest because they can serve as nano "tapes" which can immobilize desired nanoparticles, viruses, and proteins. As such, they can be used in drug delivery, controlled nanoparticle synthesis, and nanoparticle recycling. The nanoparticle-coated polymer single crystals have a sandwich structure; they might find applications in electromechanical systems and microfluidic applications.

Two of the most common nanoparticles are titania (TiO_2) and silica (SiO_2). Rashwan and Sereda (*22*) report on modifications of their surfaces with organic and inorganic groups that significantly improve the utility of these materials as ingredients in sunscreens and toothpastes. Thus, functionalization of titania and silica nanoparticles improves their adhesion to human dentin, with possible applications for treating tooth hypersensitivity by occlusion of dentin tubules and carriers of remineralizing and other active components. Moreover, the organic and inorganic modifiers can also suppress photodegradation of caffeic acid and facilitate the use of the modified silica and titania in sunscreen formulations.

High tech and energy devices often require the use of rare earth elements (REE). With increasing demand for these devices, extraction and sequestration of REE have become critical needs. Florek et al. (*23*) describe their development of nano-structured organosilica hybrid materials with chemically bonded selective ligands. These materials were found to be superior in selectively extracting REE in the presence of other competing elements, compared to commercial resins. These materials are reusable and may perhaps be utilized in environmental remediation as well.

In their chapter, Liu et al. (*24*) review recent advances in gas separation membranes by considering the materials involved, membrane fabrication, membrane modules, and industrial applications. UOP has successfully commercialized Separex™ Flux+ and Select membrane elements for natural gas upgrading and other gas separations. These new high performance membranes have a nanometer-scale selective layer, which is considered critical to the future success of membrane gas separation technologies.

Opportunities in Bio/Medicine

In his chapter, Mirkin (*12*) summarizes his work in spherical nucleic acids (SNAs), nanostructures that show extraordinary promise in biology and medicine. Typically, SNAs are synthesized by immobilizing oligonucleotides (functionalized with alkylthiol end groups) on the surface of spherical gold nanoparticles in a highly oriented and densely packed fashion. This technology is the basis for 1,800 different products on the market, including numerous medical diagnostic and therapeutic applications.

Nath (*25*) is also using nanoparticles for biomedical applications; he describes his research, which focuses on multifunctional magnetic nanomaterials, especially Au-Fe_3O_4 nanostructures. Their multiple functionalities can be utilized for diagnosis, therapeutics, cell targeting and sorting, and drug delivery. He has shown that these particles can be functionalized with bioactive molecules. In addition, the catalytic use of these nanostructures for water oxidation with multifunctional iron oxides as well as metal selenide nanostructures is discussed.

In a different approach, Neoh et al. (*26*) developed different types of soft nanoparticles as drug carriers to address the challenges encountered in intravesical chemotherapy for bladder cancer, which she details. These nanocarriers are prepared from non-cytotoxic mucoadhesive materials to promote interaction with the urothelium in order to provide sustained release of the drug locally and increase drug residence time in the bladder. With advances in nanotechnology, mucoadhesive nanocarriers can be designed to carry different types of drugs and other agents for intravesical therapy.

In their chapter, Hu and Gu (*27*) review the promising technique of utilizing cell membrane-coated nanoparticles for drug delivery and cancer treatment. The cell membrane-coated nanoparticulate drug delivery system has been extensively studied for prolonged *in vivo* circulation time, reticuloendothelial system (RES) evasion, active tumor targeting, and cancer vaccination. Nevertheless, the resulting cell membrane-coated nanoparticles need to match the patients based on their cell types to minimize the immunogenicity.

Madiyar and Li (*28*) summarize their studies of the development of a nanostructured dielectrophoresis (DEP) device for the capture of bacterial cells and virus particles. A high magnitude nonuniform electric field was produced in a microfluidic channel utilizing a nanoelectrode array made of vertically aligned carbon nanofibers versus a macroscopic indium tin oxide counter electrode in a "points-and-lid" configuration. The DEP capture was found to be fully reversible for two types of microbes including *E. coli* bacterial cells (~1000 - 2000 nm in size) and virus particles (~80 - 200 nm in size), when an AC voltage (100 Hz to 1 MHz) was turned on and off. This technique can be potentially utilized as a fast sample preparation module in a microfluidic chip to capture, separate, and concentrate microbes while analyzing small volume of dilute samples as a part of a portable detection system for field applications.

Philbert (*32*) provides a review of nanomedical drugs and devices that are already approved or are in various stages of clinical trial. As indicated in his chapter, a large number of techniques in nanotechnology are being employed for nanomedicine, e.g., nano-sized colloids, dendrimers, photoswitches, zinc finger proteins, gold nanoparticles, gold nanoshells, quantum dots, iron oxide nanoparticles, and polymer nanoparticles.

Opportunities in Food and Agriculture

In their chapter, Cheng et al. (*29*) review many R&D studies involving nanotechnology in agriculture. For example, in soil management, applications reported include nanofertilizers, soil binders, water retention aids, and nutrient monitors. In plants, nanotech methods deliver DNA to plant cells, enhance nutrient absorption, detect plant pathogens, and regulate plant hormones. In animals, nanocapsules deliver vaccines and improve delivery of nutrients. Numerous postharvest applications are reported, including the generation of nanocellulose from agriculture wastes, nanocomposites, silk, biochar, and nanosilver, among many others.

Likewise in the food-related area, nanotechnology is actively explored (for example) in food processing, packaging, nanocarrier systems for nutrients and supplements, nanosized food and feed additives, nanocoatings on food contact surfaces, nanosensors for food labeling, and water decontamination (*41*). Many of the products are now commercialized. Although this area is not included in this book, it is amply reviewed in the recent literature (*41–46*).

Challenges: Risks and Regulations

Despite the rapid developments and escalating publications in this field, there remain a number of challenges in product development and commercialization. The key issues are the health, safety and environmental aspects of some nanomaterials and their regulatory status. As pointed out by Medley (*30*), the responsible utilization and adoption of new technology will be impacted either favorably or unfavorably by the regulatory requirements for its oversight and further development. These regulatory requirements can either provide a platform which facilitates decision-making and responsible technology adoption or create unnecessary barriers to innovation and technology utilization. Yet, regulations can assist in identifying potential risks while avoiding unnecessary data generation, time delays and increased costs. The issues of risk and regulation are covered in detail in the following six chapters.

According to Morris (*31*), the Office of Pollution Prevention and Toxics of the U.S. Environmental Protection Agency (EPA) has over 10 years of experience dealing with the regulation of manufactured nanomaterials. As of November 2014, the EPA had made regulatory decisions on over 160 nanomaterials for which it had received premanufacture notification. In his chapter, Morris discusses the lessons the EPA has learned in evaluating nanomaterials with respect to environmental sustainability across the life cycle and what challenges remain in developing and applying scientific information to determine the safety of manufactured nanomaterials.

In his chapter, Philbert (*32*) also points out the Rule of Five that has been successfully used to predict the safety/toxicity of low molecular weight therapeutics developed for oral administration; yet the majority of nanomaterials in existence violate at least 2 or 3 of these rules. In contrast, he cites the work by Scott McNeil et al. at NIH's National Cancer Institute, which uncovered a series of factors that render nano-formulations safe for use. As the field of nanomedicine matures, Philbert suggests manufacturers need to be vigilant and careful in the design and deployment of this useful class of novel materials.

International governance bodies are evolving to address the benefits of nanotechnologies while seeking to manage their potential risks to human health and the environment through a variety of voluntary, standard-setting, regulatory, statutory, and related governance platforms. Bergeson (*33*) surveys emerging governance approaches with particular emphasis on the need for regulatory measures in targeted areas to ensure the integrity of core governance principles and provide some measure of commercial predictability.

In his chapter, Medley (*30*) proposes an integrated approach in order to balance appropriate regulatory oversight versus technology adoption. This approach was used at DuPont and comprised the development of a nano risk framework for responsible development, a successful OECD collaborative effort to address scientific uncertainty, and the U.S. government's issuance of science-risk-based regulatory policies and principles. He emphasizes that to be successful, this approach must be a shared responsibility among all stakeholders.

In their chapter, Murashov and Howard (*34*) discuss the proactive risk mitigation for nanotechnology workplaces. In this framework, risk is first anticipated, then recognized, and evaluated. Appropriate levels of controls are applied and their effectiveness is confirmed. This cycle is repeated as necessary to reduce risk to an acceptable level and to capture on-going changes in the workplace. Through the cooperation of all the participants involved, this framework can be used in the safe introduction of nanotechnology into our everyday lives.

Finally, Lin (*35*) in his chapter addresses the term "responsible development" that is often used in nanotechnology R&D. This term may be applied in three ways: 1) ethical conduct, e.g., avoiding conflicts of interests and treating research subjects with respect; 2) duty to account for a technology's positive and negative effects, especially its health, safety, and environmental consequences; 3) public engagement into research and examination of the R&D work in light of public viewpoints and concerns. The last item, in particular, is challenging to implement in practical and concrete ways.

Acknowledgments

Thanks are due to the authors of this book for contributing chapters. Mention of trade names or commercial products in this publication is solely for the purpose of providing specific information and does not imply recommendation or endorsement by the U.S. Department of Agriculture. USDA is an equal opportunity provider and employer.

References

1. *OECD Working Party on Nanotechnology (WPN): Vision Statement 2010*; http://www.oecd.org/sti/nano/oecdworkingpartyonnanotechnology wpnvisionstatement.htm (accessed 3/24/16).
2. Ramsden, J. *Applied Nanotechnology: The Conversion of Research Results to Products*, 2nd ed.; Elsevier: Amsterdam, 2014.
3. Sparks, S. *Nanotechnology: Business Applications and Commercialization*; CRC Press: Boca Raton, 2012.
4. Helwegen, W.; Escoffier, L. *Nanotechnology Commercialization for Managers and Scientists*; Pan Stanford Publishing: Singapore, 2012.
5. Brune, H., Ernst, H.; Grunwald, A.; Grünwald, W.; Hofmann, H.; Krug, H.; Janich, P.; Mayor, M.; Rathgeber, W.; Schmid, G.; Simon, U.; Vogel, V.; Wyrwa, D. *Nanotechnology: Assessment and Perspectives*; Springer: Berlin, 2012.
6. *Commercializing Micro-Nanotechnology Products*; Tolfree, D., Jackson, M. J., Eds.; CRC Press: Boca Raton, FL, 2008.
7. Kozhukharov, V.; Machkova, M. Nanomaterials and Nanotechnology: European initiatives, Status and Strategy (Review). *J. Chem. Technol. Metal* **2013**, *48*, 3–11.
8. Maine, E. Scientist-Entrepreneurs as the Catalysts of Nanotechnology Commercialization. *Rev. Nanosci. Nanotechnol.* **2013**, *2*, 301–308.
9. Lo, C.; Wang, C.; Chien, P.; Hung, C. An empirical study of commercialization performance on nanoproducts. *Technovation* **2012**, *32*, 168–178.
10. Hobson, D. W. Commercialization of nanotechnology. *Wiley Interdiscip. Rev.: Nanomed. Nanobiotechnol.* **2009**, *1*, 189–202.
11. The Energy and Resources Institute (TERI). *Review of international nanotechnology developments and policy concerns*; 2009; http://www.teriin.org/nano-uploads/D1_Review_of_international_ nanotechnology_developments_and.pdf (accessed 3/24/16)
12. Mirkin, C. Nanotechnology at Northwestern University: Delivering on the Promise. *Nanotechnology: Delivering on the Promise Volume 1*; ACS Symposium Series 1220; American Chemical Society: Washington, DC, 2016; Chapter 2.
13. Fadel, T. R.; Meador, M. A. The Role of Chemical Sciences in the National Nanotechnology Initiative: Accomplishments and Future Direction. *Nanotechnology: Delivering on the Promise Volume 1*; ACS Symposium

Series 1220; American Chemical Society: Washington, DC, 2016; Chapter 3.
14. Roco, M. C. Building Foundational Knowledge and Infrastructure for Nanotechnology: 2000-2030. *Nanotechnology: Delivering on the Promise Volume 1*; ACS Symposium Series 1220; American Chemical Society: Washington, DC, 2016; Chapter 4.
15. Sloter, L. Nanomaterials and nanomanufacturing with an emphasis on national security. *Nanotechnology: Delivering on the Promise Volume 1*; ACS Symposium Series 1220; American Chemical Society: Washington, DC, 2016; Chapter 5.
16. Henderson, L. National Institutes of Health: An Introduction to Nanotechnology Funded Research in Biology and Medicine. *Nanotechnology: Delivering on the Promise Volume 1*; ACS Symposium Series 1220; American Chemical Society: Washington, DC, 2016; Chapter 6.
17. Liehr, M. University Roles in Economic Development. *Nanotechnology: Delivering on the Promise Volume 1*; ACS Symposium Series 1220; American Chemical Society: Washington, DC, 2016; Chapter 7.
18. Lewis, N. S.. Progress Towards a Synergistically Integrated, Scalable Solar Fuels Generator. *Nanotechnology: Delivering on the Promise Volume 2*; ACS Symposium Series 1224; American Chemical Society: Washington, DC, 2016; Chapter 1.
19. Rajeeva, B. B.; Zheng, Y. Molecular Plasmonics: From Molecular-Scale Measurements and Control to Applications. *Nanotechnology: Delivering on the Promise Volume 2*; ACS Symposium Series 1224; American Chemical Society: Washington, DC, 2016; Chapter 2.
20. Iocozzia, J. A.; Lin, Z. Crafting multidimensional nanocomposites: Functional materials for application in energy conversion, energy storage and optoelectronics. *Nanotechnology: Delivering on the Promise Volume 2*; ACS Symposium Series 1224; American Chemical Society: Washington, DC, 2016; Chapter 3.
21. Qi, H.; Mei, S.; Zhou, T.; Dong, B.; Li, C. Y. Nanoparticle-Decorated Polymer Single Crystals for Nanoscale Materials. *Nanotechnology: Delivering on the Promise Volume 2*; ACS Symposium Series 1224; American Chemical Society: Washington, DC, 2016; Chapter 4.
22. Rashwan, K.; Sereda, G. Applications of Nanoparticles through Surface Functionalization. *Nanotechnology: Delivering on the Promise Volume 2*; ACS Symposium Series 1224; American Chemical Society: Washington, DC, 2016; Chapter 5.
23. Florek, J.; Larivière, D.; Kleitz, F. Nanostructured Organosilica Hybrids as Highly Efficient and Regenerable Sorbents for Rare Earth Extraction. *Nanotechnology: Delivering on the Promise Volume 2*; ACS Symposium Series 1224; American Chemical Society: Washington, DC, 2016; Chapter 6.
24. Liu, C.; Greer, D. W.; O'Leary, B. W. Advanced Materials and Membranes for Gas Separations. *Nanotechnology: Delivering on the Promise Volume*

2; ACS Symposium Series 1224; American Chemical Society: Washington, DC, 2016; Chapter 7.
25. Nath, M. Multifunctional Magnetic Nanomaterials for Diverse Applications. *Nanotechnology: Delivering on the Promise Volume 2*; ACS Symposium Series 1224; American Chemical Society: Washington, DC, 2016; Chapter 8.
26. Neoh, K. G.; Lu, S.; Kang, E.-T.; Mahendran, R.; Chiong, E. Tailoring Soft Nanoparticles for Potential Application as Drug Carriers in Bladder Cancer Chemotherapy. *Nanotechnology: Delivering on the Promise Volume 2*; ACS Symposium Series 1224; American Chemical Society: Washington, DC, 2016; Chapter 9.
27. Hu, Q.; Gu, Z. Cell Membrane-Mediated Anticancer Drug Delivery. *Nanotechnology: Delivering on the Promise Volume 2*; ACS Symposium Series 1224; American Chemical Society: Washington, DC, 2016; Chapter 10.
28. Madiyar, F. R.; Li, J. Nanoelectrode Array Based Devices for Electrical Capture of Microbes Using Dielectrophoresis. *Nanotechnology: Delivering on the Promise Volume 2*; ACS Symposium Series 1224; American Chemical Society: Washington, DC, 2016; Chapter 11.
29. Cheng, H. N.; Klasson, K. T.; Asakura, T.; Wu, Q. Nanotechnology in Agriculture. *Nanotechnology: Delivering on the Promise Volume 2*; ACS Symposium Series 1224; American Chemical Society: Washington, DC, 2016; Chapter 12.
30. Medley, T. L. Regulation: Facilitating advancement or serving as a barrier – a shared responsibility. *Nanotechnology: Delivering on the Promise Volume 1*; ACS Symposium Series 1220; American Chemical Society: Washington, DC, 2016; Chapter 8.
31. Morris, J. T. Sustainability and Life Cycle Issues. *Nanotechnology: Delivering on the Promise Volume 1*; ACS Symposium Series 1220; American Chemical Society: Washington, DC, 2016; Chapter 9.
32. Philbert, M. Nanomaterials: Promise in Balance with Safety. *Nanotechnology: Delivering on the Promise Volume 1*; ACS Symposium Series 1220; American Chemical Society: Washington, DC, 2016; Chapter 10.
33. Bergeson, L. L. Opportunities and Challenges for Health, Safety, and the Environment: The Regulatory Void? *Nanotechnology: Delivering on the Promise Volume 1*; ACS Symposium Series 1220; American Chemical Society: Washington, DC, 2016; Chapter 11.
34. Murashov, V.; Howard, J. Adaptive Governance for the Nanotechnology Workplace. *Nanotechnology: Delivering on the Promise Volume 1*; ACS Symposium Series 1220; American Chemical Society: Washington, DC, 2016; Chapter 12.
35. Lin, A. C. What is responsible development of nanotechnology? *Nanotechnology: Delivering on the Promise Volume 1*; ACS Symposium Series 1220; American Chemical Society: Washington, DC, 2016; Chapter 13.

36. National Nanotechnology Initiative. http://www.nano.gov/about-nni (accessed 3/24/16)
37. Whitman, L. J. Lecture given at the ACS National Meeting in Denver, March 2015 during the symposium on "Nanotechnology: Delivering on the Promise."
38. President's Council of Advisors on Science and Technology. Report to the President and Congress on the 5th Assessment of NNI; www.whitehouse.gov/sites/default/files/microsites/ostp/PCAST/pcast_fifth_nni_review_oct2014_final.pdf (accessed 3/24/16)
39. Harper, T. *Global funding of Nanotechnologies and its Impact*; http://cientifica.com/wp-content/uploads/downloads/2011/07/Global-Nanotechnology-Funding-Report-2011.pdf (accessed 3/25/16)
40. Milanović, V.; Bučalina, A. *Position of the Countries in Nanotechnology and Global Competitiveness*; http://www.management.fon.rs/management/e_management_68_english_08.pdf (accessed 3/25/16)
41. Anon. *Food Nanotechnology*; http://www.nanowerk.com/nanotechnology-in-food.php (accessed 3/27/16)
42. Sozer, N.; Kokini, J. L. Nanotechnology and its applications in the food sector. *Trends Biotechnol.* **2009**, *27*, 82–89.
43. Duncan, T. V. Applications of nanotechnology in food packaging and food safety: Barrier materials, antimicrobials and sensors. *J. Colloid Interface Sci.* **2011**, *363*, 1–24.
44. Chaudhry, Q.; Castle, L. Food applications of nanotechnologies: An overview of opportunities and challenges for developing countries. *Trends Food Sci. Technol.* **2011**, *22*, 595–603.
45. Misra, A. N.; Misra, M.; Singh, R. Nanotechnology in Agriculture and Food Industry. *Int. J. Pure Appl. Sci. Technol.* **2013**, *16*, 1–9.
46. Pray, L.; Yaktine, A. *Nanotechnology in Food Products*; National Academies Press (U.S.): Washington, DC, 2009.

Managing Nanotechnology: R&D

Chapter 2

Nanotechnology at Northwestern University: Delivering on the Promise

Chad A. Mirkin*

Department of Chemistry and International Institute for Nanotechnology, Northwestern University, 2145 Sheridan Road, Evanston, Illinois 60208, United States
***E-mail: chadnano@northwestern.edu.**

The dawn of the 20th century saw the birth of the airplane, the radio, and the theory of relativity—three technological and scientific developments that changed the world. Now a hundred years later, nanotechnology is similarly transforming the world and holds the promise to solve some of the world's most pressing problems in areas as diverse as medicine, information technology, energy, and the environment. Nanotechnology is especially poised to make a large impact on the fields of biology and medicine. Nanomaterials are the ideal size to efficiently interact with biological structures and thus useful for both *in vivo* and *in vitro* biomedical research and medical applications. Indeed, these methods have the potential to change the way we study and treat some of the world's most debilitating diseases, including cancer, cardiovascular disease, and Alzheimer's disease. This article introduces the readers to the International Institute for Nanotechnology at Northwestern University and also summarizes some of the author's own work in spherical nucleic acids (SNAs), nanostructures that have shown extraordinary promise in biology and medicine.

International Institute for Nanotechnology

The International Institute for Nanotechnology (IIN) at Northwestern University is an umbrella organization that unifies endeavors in nanoscience and technology by researchers on the Evanston and Chicago campuses of Northwestern, Northwestern's Feinberg Medical School, and neighboring Chicago-area research institutions, including Argonne National Laboratory (ANL). There are many strong collaborations among the researchers at these facilities. The IIN is building a collective infrastructure to pull together teams that can use nanoscience and technology to address scientific problems in almost any topical area.

The work at this institute has led to a combined R&D throughput of over $800 million since its founding in 2000. With the support of its current president, Dr. Morton Shapiro, and previous ones, including Dr. Henry Bienen, and others, Northwestern's IIN has created global partnerships in 18 different countries, established 36 different research centers, created over 100 corporate partnerships, and seen the participation of over 750 graduate students and postdoctoral fellows as well as over 190 faculty members across 25 departments.

The International Institute for Nanotechnology focuses on multiple topical areas:

- Materials and devices;
- Environmental science;
- Information technology;
- Homeland security;
- Food and water safety;
- Energy;
- Transportation;
- Medicine.

Much of the IIN's operations focuses on catalyzing and supporting world-class interdisciplinary nanoscience research to address the world's most pressing problems. The teams that congregate around the different areas focus their attention on main technological drivers. This focus allows the teams to create new scientific and technological innovations that benefit society. More information on the IIN can be found on its website (*1*).

Small Business Evaluation and Entrepreneur's (SBEE) Program

The IIN has a partnership with Northwestern University's Kellogg School of Management - the SBEE program (*2*). As a part of this program, scientific researchers who are developing new technologies work with teams of Kellogg students. This partnership allows scientists, engineers, and medical doctors to present new discoveries that could fuel the creation of start-up companies and commercial products. These ideas are vetted by business students, who assess the potential business opportunities related to the ideas. The careful consideration of market goals at the early stages of innovation and an understanding the market applications of a technology can influence research directions and priorities. The business students and researchers put together a finely tuned business plan that makes a case for whether to move forward with commercialization or development or not. This information can be brought to potential investors, allowing the ideas to receive a lot more attention than other ideas that have not been market-tested.

This process has helped enable the start-up of 20 different companies and changed the way Northwestern does business in the Chicago area. Prior to the establishment of this program, there was less than $50 million investment in hardcore technology, focusing on materials research and biotechnology. Now, over $700 million in venture capital has been garnered to promote technological investment. Within the 20 companies, there have been major product successes, and some of these companies have been acquired by larger entities. Overall, this process has given these companies a much greater chance at success.

Convergence of Nanoscience and Nanotechnology with Medicine

Being able to reliably make new nanomaterials allows us to be able to develop high-sensitivity and high-selectivity molecular diagnostic tools and therapeutic agents, changing how we study, track, and treat disease. These discoveries have led to new *in vivo* imaging materials, which provide better contrast and offer theranostic capabilities; tools for genetically manipulating cells and making intracellular measurements; and novel approaches to drug delivery, gene regulation therapy, and immune-engineering. Nanotechnology-based drug delivery systems can be utilized to effectively treat the disease at a particular location in the body with fewer side effects as compared to traditional approaches. Northwestern has developed a series of technologies through the Centers of Cancer Nanotechnology Excellence (CCNE), funded by the National Cancer Institute (NCI) (*3*).

Spherical Nucleic Acids (SNAs)

Arguably, one of the most important discoveries of the last century was the determination of the structure of nucleic acids, including DNA - the blueprint of life. It gave humans the basis to create all sorts of new and powerful medical diagnostic tools and types of therapies based upon the use of genetic sequences and the knowledge of how they interface with biological systems. It gave rise to the human genome project.

In 1996, a new, three-dimensional form of nucleic acid was synthesized and introduced by researchers at Northwestern University - spherical nucleic acids (SNAs). This structure organizes perhaps the most important molecule ever synthesized by chemists - DNA – into a new form, imparting upon the conjugate novel chemical and physical properties that distinguish it from all other forms of nucleic acids (including those that make them useful in biology and medicine). Typically, SNAs are synthesized by immobilizing oligonucleotides (functionalized with alkylthiol end groups) on the surface of spherical gold nanoparticles in a highly oriented and densely packed fashion. (Figure 1). One can control the composition of this construct based by toggling the size of the nanoparticle and its composition and the length, type, and sequence of oligonucleotides used (*4*).

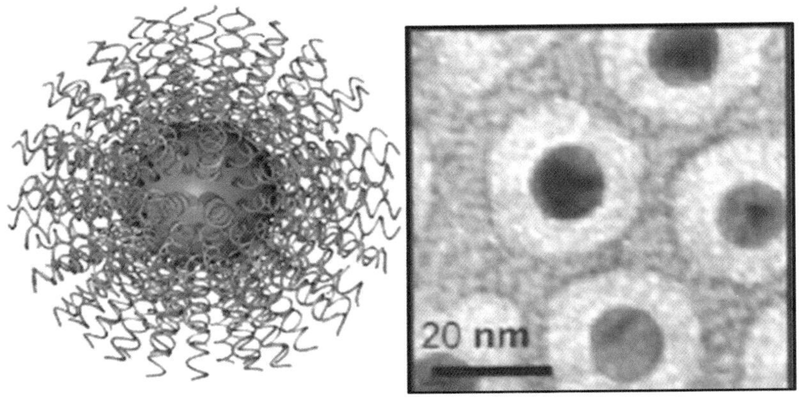

Figure 1. Schematic of the structure of a SNA (left, from Ref. (4)). Transmission electron microscopy (TEM) image of SNAs with gold nanoparticle cores and 40-mer oligonucleotide shells (with uranyl acetate staining) (right Courtesy of author).

If one compares SNAs to linear nucleic acids on a sequence-for-sequence basis, they are very different. SNAs will bind ~100 times more tightly to complementary nucleic acids than linear nucleic acids of the same sequence due to the entropic and ordering effects that come from arranging the oligonucleotides on the surface of a particle. SNAs exhibit very narrow, cooperative melting transitions that look like first-order phase transitions; linear oligonucleotides of the same sequence exhibit broader transitions that occur at lower temperatures. SNAs can be constructed from a variety of different types of nanoparticles (including those of gold, silver, iron oxide, and silica); the core can even be removed to create a SNA structure with a hollow core. Many of the unique properties of SNAs stem from the three-dimensional architecture of the nucleic acids. SNAs will efficiently enter cells without the use of secondary, and often toxic, transfection agents, while linear nucleic acids will not. Once inside, they persist about 10 times longer because their structure protects them from degradation. This is important in terms of developing therapeutics and diagnostics. SNAs also elicit a very minimal immune response, unless purposely designed to engage the immune system as a means of therapeutic action.

SNAs are changing paradigms in the field of biology. Cell membranes are negatively charged, and the lore prior to the discovery and development of SNAs was that nucleic acids would not effectively enter cells without the use of materials such as positively charged transfection agents or viruses. However, three-dimensional SNAs have been found to not only enter cells (over 60 types tested to date, except mature red blood cells), but enter them better than almost any other material known to man (explored via flow cytometry, among other techniques). Cell surfaces present class-A scavenger receptors that have a much higher affinity for SNAs than linear nucleic acids. SNAs are taken into cells via caveolin-mediated endocytosis, and when the endosomes created during this process break, the SNAs are released into the cytoplasm.

Commercial Forms and Uses of Spherical Nucleic Acids

President Barak Obama came to Northwestern and gave a speech to rally support for the IIN (*5*). In his speech, he said "If we want to make and sell the best products, we have to invest in the best ideas like we do at Northwestern. Your nanotechnology doesn't just conduct groundbreaking research. That research has spun off 20 startups and more than 1,800 products. That means jobs!" Many of the companies and products he was referring to involve SNAs. There are 1,800 different products based on SNAs on the market, and there will be many more in the years to come.

SNAs are used in medical diagnostic tools (i.e., as labels in the FDA-cleared Verigene System™ sold by Nanosphere (Northbrook, IL)) (*6*). Ten FDA-cleared panel assays are based upon the use of these particles as probes. David Pogue, a New York Times Science writer, now working for Google, was doing an episode of NOVA called "Making Stuff Smaller." He drove to Nanosphere and got involved himself, showing the power of the technology in real time. These systems are working, and they are available all over the world. As he describes in a NOVA episode (*7*),

"Chad Mirkin has developed a technology that harnesses the unique properties of gold and silver nanoparticles to test for genetic variations in patients. Sequencing DNA is expensive and time consuming, but Chad's revolutionary test takes less than two hours. I offered to bleed for science, to see how it works. First, a technician extracts pure DNA from my blood sample. He then loads it into a small disposable cartridge and inserts it into a machine for testing. This test can actually read the letters of my DNA, and using gold nanoparticles it flags variations that might make me unusually sensitive to particular drugs, or even mutations that signal heightened risk for a disease. Less than two hours after drawing my blood, the results are in. It turns out that the test has some interesting news for me about my sensitivity to a blood thinning drug called Warfarin or Coumadin commonly prescribed to stroke and cardiac patients. It's a potentially life-saving drug, but if the dosage is wrong it can cause fatal bleeding. The Nanosphere test gave me crucial information, quickly enough to save my life if I had been ill. It's a diagnostic tool. The next goal is to fight illness in the body at the same tiny scale."

There are many applications for this process, and it can be used to target a wide variety of ailments. SNAs are currently being pursued aggressively to monitor the progression of an infection to sepsis or septic shock.

Further, SNAs are the basis for NanoFlare technology, commercialized under the trade name SmartFlares™. Merck Millipore, in partnership with AuraSense (Skokie, IL), a company spun-out of Northwestern based on research initiated there, has now commercialized over 1,800 different kinds of SmartFlares for research use. SmartFlares are SNAs hybridized with short fluorophore-labeled "flare" sequences complementary to a target mRNA of interest. When a SmartFlare encounters the target mRNA, the short "flare" sequence is displaced as the mRNA binds to the SNA structure. The fluorophore is no longer quenched by its close proximity to the gold particle, and therefore a large increase in fluorescence is observed. In short, SmartFlares are designed to "turn on" intracellularly when they encounter a target mRNA sequence of interest. SmartFlares can be used in live cells, including circulating tumor cells (CTCs) and stem cells, and represent the only way of measuring the genetic content of living cells. Depending on sequence design, SmartFlares can, for instance, be used to differentiate cancer cells from healthy cells by targeting a gene that is overexpressed in cancer cells (*4, 8, 9*). SmartFlares also offer the opportunity for researchers to sort and study how cells with different genetic make-ups respond to different types of therapeutics. SmartFlares have important implications in stem cell RNA characterization, gene regulation monitoring in real time, live cell sorting via flow cytometry, and high-throughput drug screening. In addition, by changing the structure of the NanoFlare such that the "flare" sequence becomes associated with the target mRNA, the position of the mRNA in the live cell can be tracked (sticky-flares). This system has important implications on studies in developmental biology, cell biology, and neurobiology.

In addition, SNAs have many and diverse uses in the therapeutic arena – particularly in gene regulation and immunomodulation. SNAs can be utilized as:

- Antisense structures that can soak up target mRNA and down-regulate protein production (including those that cause disease);
- Agents that up- or down-regulate mRNA production;
- siRNA and miRNA delivery systems;
- Immuno-modulatory agents that enter cells and engage toll-like receptors (TLRs) to stimulate or suppress an immune response.

The vast majority of nucleic acid-based drug delivery systems used today target the liver. Spherical nucleic acids, because of their unique biodistribution properties and ability to be locally applied, can be targeted to many different areas of the body, including the brain and the skin.

References

1. International Institute for Nanotechnology; http://www.iinano.org/ (accessed 1/22/16).
2. Small Business Evaluation & Entrepreneur's (SBEE) Program at Northwestern University; http://www.iinano.org/small-business-partnership (accessed 1/22/16).

3. Northwestern Cancer Center for Nanotechnology Excellence; http://cancer.northwestern.edu/research/research_programs/ccne/index.cfm (accessed 1/22/16).
4. Cutler, J. I.; Auyeung, E; Mirkin, C. A. Spherical Nucleic Acids. *J. Am. Chem. Soc.* **2012**, *134*, 1376–1391.
5. Remarks by President B. Obama on the economy -- Northwestern University, October 2, 2014. https://www.whitehouse.gov/the-press-office/2014/10/02/remarks-president-economy-northwestern-university.
6. Nanosphere, Inc. www.nanosphere.us/ (accessed 1/22/16).
7. PBS on-line (Corporation for Public Broadcasting): NOVA. http://www.pbs.org/wgbh/nova/tech/making-stuff.html (accessed 1/22/16).
8. Giljohann, D.A.; Seferos, D.S.; Daniel, W.L.; Massich, M.D.; Patel, P.S.; Mirkin, C.A. Gold Nanoparticles for Biology and Medicine. *Angew. Chem., Int. Ed.* **2010**, *49*, 3280–3294.
9. Website for Mirkin Research Group at Northwestern University. http://mirkin-group.northwestern.edu/research/spherical-nucleic-acids-snas/ (accessed 1/22/16).

Chapter 3

The Role of Chemical Sciences in the National Nanotechnology Initiative: Accomplishments and Future Direction

Tarek R. Fadel*,[1] and Michael A. Meador[2]

[1]The International Technology Research Institute, 518 S. Camp Meade Road, Suite 6, Linthicum, Maryland 21090, United States
[2]The National Nanotechnology Coordination Office, 4201 Wilson Boulevard, Suite 405, Arlington, Virginia 22230, United States
***E-mail: mmeador@nnco.nano.gov.**

As a core component of nanotechnology, chemical sciences allow for the precise crafting of new structures in a wide range of commercial applications. Fifteen years since President Clinton announced the National Nanotechnology Initiative (NNI), over twenty billion dollars have been invested in fundamental and applied nanotechnology research and development; world-class characterization, testing, and fabrication facilities; education and workforce development; and efforts directed at understanding and controlling the environmental, health, and safety characteristics of engineered nanomaterials and consumer products derived from them. The NNI has not just evolved in structure and emphasis; it now is transitioning towards a critical inflection point where greater focus is being placed on the rapid commercialization of nanotechnologies. This chapter will briefly review the structure and history of the NNI, highlight a few examples of advancement in nanoscience pertaining to Clinton's grand challenges, and provide a synopsis of what the future may hold for the initiative and nanoscience in the U.S.

Introduction

When chemists hear of nanotechnology, their reaction could be a simple shrug. After all, the nanoscale (or a billionth of a meter) is an intimate dimension of chemistry. As Whitesides plainly puts it, "chemistry *is* (and has always been) the ultimate nanotechnology: Chemists *make* new forms of matter (and they are really the only scientists to do so routinely) by joining atoms and groups of atoms together with bonds" (*1*). A special emphasis, however, should be on the word "make." The process of "making" is indeed an essential step of nanoscience, where new forms of matter are engineered to exhibit some interesting properties because of certain phenomena that occur at the 1 nm to 100 nm range (*2*). More importantly, these properties may differ in significant ways from the properties of bulk materials and single atoms or molecules (*3*).

Nanotechnology has now evolved as an enabling discipline crossing more traditional fields within physical, life, and social sciences and engineering (*4*). As a transformative and potentially pervasive technology, it has catalyzed the development of new materials and processes that can collect and store energy (*5*), reduce the weight and improve the performance of structures (*6*), sense chemicals (*7*), enable lifesaving drugs (*8*), and enhance computational devices in both incremental and paradigm-shifting ways (*9*).

Chemistry and the chemical sciences have played a central role in nanotechnology discovery and commercialization. A quick review of the literature could provide some preliminary insights on the contribution of chemical sciences towards nanotechnology-related publications and patents. Using a dataset that identifies nanotechnology papers and patents by means of a Boolean search in Scopus [nanotechnology OR nanoscience OR nanoparticle OR nanoparticulate OR nanomaterial], over 469,000 Scopus publications were mined for the period of 2000-2014, as well as over 10,000 patents. When limiting the subject to "chemistry" or "chemical engineering" in Scopus, the search yielded about 214,000 publications and 2900 patents for the same period, so about 46% and 27%, respectively. As a reference, narrowing publications in the subject area of "material sciences" yielded a top contribution of 47% and 31% towards publications and patents, respectively, while selecting the subject of "medicine" yielded 9% and less than 1% for publications and patents, respectively. Overall, chemical and material sciences were found to be among the most prominent subject areas contributing to nanotechnology publications and patents.

In the U.S., the National Nanotechnology Initiative (NNI) is the Federal Government's interagency activity that coordinates research and development (R&D) and fosters communication and collaborative activities in nanoscale science, engineering, and technology (*10*). The NNI defines nanotechnology as "the understanding and control of matter at dimensions between approximately 1 and 100 nanometers, where unique phenomena enable novel applications" (*10*). Fifteen years have gone by since former President Clinton announced this major initiative in a landmark speech at the California Institute of Technology in Pasadena (*11*). In that speech, President Clinton laid out three concrete goals for the initiative: "Imagine the possibilities: materials with ten times the strength of steel and only a small fraction of the weight—shrinking all the information housed at the Library of Congress into a device the size of a sugar cube—detecting cancerous tumors when they are only a few cells in size." He even suggested a timeframe for this tracking exercise. "Some of our research goals may take 20 or more years to achieve, but that is precisely why there is an important role for the Federal government."

Now that fifteen years have gone by, some might ask: what progress has been made towards President Clinton's goals? How has investment in nanotechnology research and development led to products that benefit society and the Nation? What does the future hold for the NNI and nanotechnology research, development and commercialization? This chapter will address these questions with an eye towards highlighting examples of where advances in the chemical sciences have contributed to the success of the NNI, and how chemistry and the chemical sciences can potentially contribute to the future directions and successes of the initiative.

A Look Back at the National Nanotechnology Initiative

About the NNI

The National Nanotechnology Initiative (NNI) is a U.S. Government research and development initiative aimed at the shared and challenging vision of "a future in which the ability to understand and control matter at the nanoscale leads to a revolution in technology and industry that benefits society" (*10*). Shortly after the NNI was announced by President Clinton in January 2000, it was joined by six agencies: Department of Defense (DOD), Department of Energy (DOE), National Aeronautics and Space Administration (NASA), National Institutes of Health (NIH), National Institute of Standards and Technology (NIST), and National Science Foundation (NSF). The total investment was about $490 million, including congressionally directed funding (*12*).

The initiative has since evolved in both structure and emphasis to a partnership of twenty Federal agencies and departments with activities in nanotechnology R&D, policy, and regulation (Figure 1a; see the list of acronyms at the end of the chapter). The investment has also grown from $464 million per year to almost $1.5 billion requested in the President's 2016 Budget, totaling more than $22 billion since the inception of the initiative (Figure 1b) (*13*). This progress has been guided by the bipartisan support of several administrations as part of an ongoing innovation strategy. The coordinated efforts of agencies participating in the NNI have indeed accelerated discovery, development, and deployment of nanotechnology to benefit agency missions in service of the broader national interest (*13*).

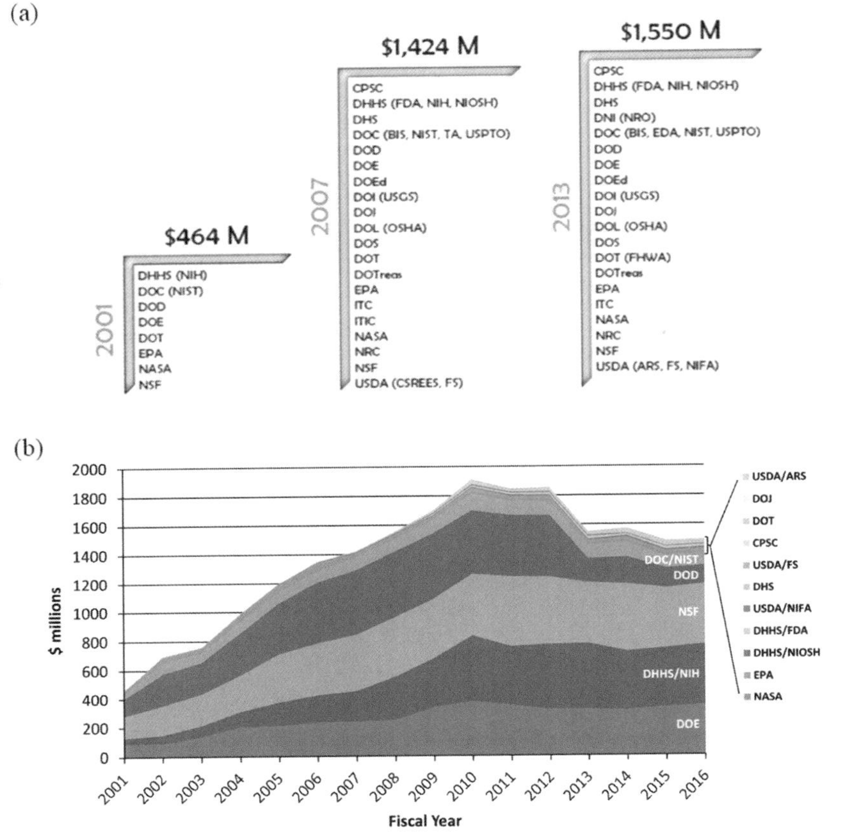

Figure 1. NNI Membership and Budgets. (a) Diagram describing evolution of agencies and budgets in the NNI from 2001 to 2013. Source: adapted from A. H. Carim, NNI R3 Workshop (2013). (b) Plot of NNI funding by agency from 2001to 2016. For 2009, budget figures do not include American Recovery and Reinvestment Act funds for DOE ($293 million), NSF ($101 million), NIST ($43 million), and NIH ($73 million); for 2015, estimated figures are based on enacted levels and may shift as operating plans are finalized. Source: NNI Supplement to the President's 2016 Budget (2015). For agency abbreviations, please see the list of acronyms at the end of the chapter.

The NNI is coordinated by a subgroup of the White House National Science and Technology Council (NSTC). In 1998, the Interagency Working Group on Nanoscale Science, Engineering and Technology (IWGN) was established to assess the potential of nanotechnology and formulate a national research and development plan (*12*). In July 2000, IWGN evolved into the subcommittee level with the creation of the Nanoscale Science, Engineering, and Technology (NSET) Subcommittee of the NSTC's Committee on Technology (Figure 2).

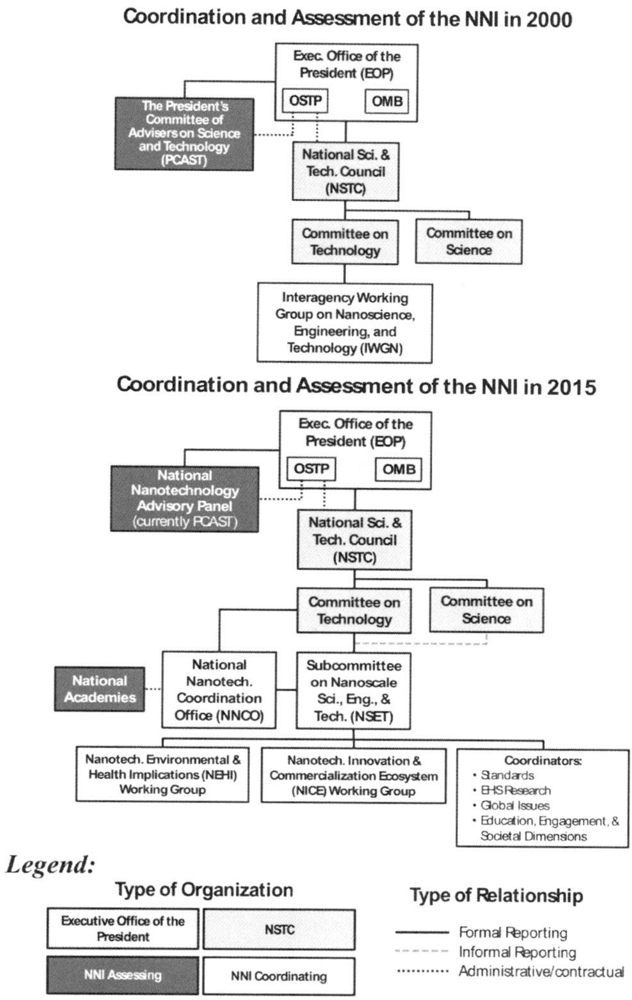

Figure 2. NNI Structure from 2000 to 2015. Sources: adapted from the IWGN "National Nanotechnology Initiative: Leading to the Next Industrial Revolution" report (2000) and the NNI Strategic Plan (2014).

Federal agencies participating in the NSET Subcommittee coordinate the establishment of goals and strategies that complement agency-specific missions and activities (*10*). The subcommittee also provides a central channel for identifying common priorities, leveraging resources, and engaging with the public. These efforts are supported by the National Nanotechnology Coordination Office (NNCO), which serves as a central point of contact for Federal nanotechnology R&D activities and provides public outreach on behalf of the NNI (*10*). For example, every fiscal year Federal agencies participating in NSET, with support from NNCO, publish a supplement to the President's budget request (*13*). This document is submitted to Congress and serves as the Annual Report for the NNI called for under the provisions of the 21st Century Nanotechnology Research and Development Act of 2003 (*14*).

"Imagine What Could Be Done"

When former President Clinton unveiled the NNI in his major science policy address at the California Institute of Technology on January 21, 2000 (*11*), he channeled excitement for the emergence of nanotechnology with the words "imagine what could be done." He also included in his speech a series of ambitious yet achievable challenges that he identified with the support of the research community and that would be easily communicated to the general public. These were "developing materials with ten times the strength of steel and only a small fraction of the weight," "storing the Library of Congress in a device the size of a sugar cube," and "detecting cancerous tumors before they are visible to the human eye." Examples from the literature can provide some insights on the progress towards these three challenges, and particularly how chemical sciences have enabled them. Inevitably, the nature of this progress (as in many other scientific disciplines) has advanced our current knowledge of basic sciences as well as revitalized our approach to physical sciences and engineering.

Materials with Ten Times the Strength of Steel and a Fraction of the Weight

Years ago, the symbolism behind a carbon nanotube space elevator (*15*) not only captured the imagination of many, but offered a window into how nanotechnology could enable a revolution in the development of ultra-high strength structural materials (*16*). Recent research on advanced materials has achieved key milestones for structural applications in the aerospace and defense sectors requiring high strength, lightweight properties (*17*). Single-walled carbon nanotubes for instance have been shown to exhibit Young's modulus of approximately one terapascal, fifty times that of steel, tensile strengths in the range of fifty to one hundred gigapascal, and specific strengths up to three-hundred times that of high-carbon steel (*18*). In practice, carbon nanotube-based sheet materials have been introduced by Nanocomp Technologies, Inc. to help protect vital components of the Juno spacecraft (*19*). Another example is the use of cost-efficient advanced nanocomposites by Lockheed Martin Corp. onboard the F-35 Joint Strike Fighter (*20*). Nanocomposite materials are also believed to be used in the fuselage of the Boeing 787 (*21*).

Early efforts in the utilization of carbon nanotubes to develop high strength materials involved their dispersion in a variety of polymers. While significant progress has been made in this area, the improvements in mechanical properties overall have been limited by agglomeration of the nanotubes as loadings above a few weight percent (*22*). A more promising approach that has gained attention over the past few years is to incorporate the nanotubes directly into fibers that could be used as drop-in replacements for more conventional fiber reinforcements in composites. Carbon nanotube fibers have been prepared using a variety of methods, including dry spinning from vertically aligned nanotube arrays (*23*) and nanotube aerogels (*24*), and wet/solution spinning from lyotropic solutions of carbon nanotubes in strong acids (*25*) and polyelectrolytes (*26*). The tensile properties of these fibers are controlled by two competing factors—Van der Waals forces that hold the nanotubes together and the low coefficient of friction of carbon nanotubes that causes them to slide against each other under tension. Improvements in the tensile strength and modulus of these fibers have been demonstrated by introducing cross-links between nanotubes using functionalization (*27*), e-beam irradiation (*28*), and a combination of these approaches (*29*).

While considerable progress has been made in developing high strength carbon nanotube fibers and utilizing them to make ultra-high strength composites (*30*), much work remains to raise the properties of these bulk materials to those of individual carbon nanotubes (*31*). Chemical sciences (including informatics) can play an important role in not only minimizing cost of production of these nanomaterials and composites, but also in engineering robust dispersion, stability, and quality processes to enable scalable production of bulk forms of these nanomaterials (*31*). Advances in synthetic chemistry (including polymer sciences) and understanding of structure-property relationships are also critical to achieve precision-crafting of such materials while maintaining strength properties at the macroscale.

Storing the Library of Congress in a Device the Size of a Sugar Cube

Our world today generates an astronomical amount of data. According to an analysis from the Economist (*32*), accumulation of data is growing at a compound annual rate of 60% and keeps on increasing. When former President Clinton referred to "storing the Library of Congress in a device the size of a sugar cube," he introduced a target for information storage that moves along a rapid current of data inflation (for reference, the Economist estimated in 2010 that all the catalogued books in the Library of Congress amounted to 15 terabytes (*32*)). In order to amass this flood of digital information, some research has focused on shrinking the spatial dimensions of devices' individual bits. However, such approaches could have limited success due to device scaling challenges (*33*). From a chemical sciences perspective, engineering new molecular approaches by altering the information storage medium holds significant promise in achieving Clinton's data storage challenge. For example, researchers from IBM and the German Center for Free-Electron Laser Science recently succeeded in storing one bit of data in as little as twelve atoms. This nanostructure was engineered by aligning two rows of six iron atoms on a surface of copper nitride at a temperature near absolute zero (*34*). Although a proof of concept at the time, the resulting storage density could be potentially about a hundred times greater than current hard disk drive technologies (about 50 to 100 Terabits per square inch).

Rapid advancement in the synthesis and sequencing of deoxyribonucleic acid (DNA) has also paved the way for novel alternatives to magnetic and optical storage mediums (*35*). Indeed, chemical engineering of complex DNA nanostructures has been demonstrated in structural biology, biocatalysis, and drug delivery-related applications (*36*), and the per-base cost of DNA sequencing has plummeted by about a hundred thousand-fold over the past decade, far outpacing Moore's law (*37*). Now, DNA-based storage media are being demonstrated by various research groups. Notably, Church and colleagues at the Harvard Medical School and the Wyss Institute of Biologically Inspired Engineering published in 2012 a strategy to encode arbitrary digital information in DNA, and outlined how each gram of DNA could store 455 exabytes of data (one exabyte is one quintillion bytes) (*38*). These researchers achieved 5.27 megabit-size coding from a book into roughly 55,000 oligonucleotides and successfully sequenced the data that was stored in the DNA medium.

Other efforts to develop new information storage medium have explored building binary data into strands of synthetic polymer. Control of monomer sequences at the molecular level allows for encryption of any desired sequence in the polymer chain, where monomers or functional side-groups could be regarded as information bit (*39*). For example, recent work by Roy and co-workers intentionally assigned zero or one values to monomers, which were then assembled in a specific order to store information into the polymer (*40*). This approach relies on the use of two successive chemoselective coupling steps, and can be easily sequenced using mass spectrometry.

Overall, further advances in nanotechnology-based information storage will occur by continuing to exploit the intersections of the chemical sciences with engineering, physics, and biology. Considerations of data security and protection of privacy should be an essential design component when developing these technologies.

Detecting Cancerous Tumors before They Are Visible to the Human Eye

Cancer continues to plague the world. According to PBS's Cancer: The Emperor of All Maladies, "more will die from cancer [in the U.S.] over the next two years than died in combat in all the wars the United States has ever fought, combined" (*41*). Although more options for the treatment of various cancers have been developed, an important issue remains with the earlier detection of tumors in patients, which is central to the success of cancer therapy (*42*). For example, more than 80% of patients diagnosed with lung cancer present a metastatic form of the disease (*43*). In early stage cancers, various biomarkers are present in the bloodstream at low concentration, as well as cancer cells from primary tumors, known as circulating tumor cells (CTCs) (*44*). In this case, advances in sensor nanotechnologies could make early detection of CTCs and tumor biomarkers a clinical reality: for example, with magnetic nanoparticles that are functionalized with an affinity ligand, or polymer nanoparticles with long circulation time (*43*).

In vitro nanotechnology-based devices for the early detection of CTCs are now being offered at the clinic. For example, Veridex's CellSearch® uses a ferrofluid reagent that consists of nanoparticles with a magnetic core surrounded by a polymeric layer and coated with target antibodies (*45*). After capture and enrichment, fluorescent reagents are added for identification and enumeration of CTCs. This system aims to provide a rapid, precise, and reproducible platform to capture CTCs of epithelial origin from whole blood and help determine patient prognosis (*46*). The CellSearch® platform was shown to detect as little as 5 CTCs per 7.5 ml of blood (*47*), and is currently being used to aid in the monitoring of patients with metastatic breast, colorectal, or prostate cancer.

An alternate method for the earlier diagnosis of cancer focuses on the detection of tumor biomarkers. For example, the Verigene™ system by Nanosphere is an automated platform based on disposable test cartridges (*48*). Samples loaded onto the cartridge are treated for nucleic acid extraction, purification, and hybridization, which is carried out using oligonucleotide-conjugated gold nanoparticles. Although not currently approved by the FDA for clinical applications in cancer, Nanosphere's development of the Verigene platform provides the capability to quantify markers indicative of certain tumors at very low concentrations using high-sensitivity protein diagnostics (femtomolar range) (*49*). Preliminary work demonstrated clinical relevance in bladder, kidney, and prostate cancers (*50*). Google Life Sciences is researching similar efforts by developing an *in vivo* nanoparticle-based sensing platform that can detect cancer biomarkers directly in the bloodstream (*51*). A magnetically-active wearable device would then attract and count the target-bound nanoparticles to monitor the progression of the disease.

A hallmark of nanotechnology-enabled drug delivery has been the capacity to exploit a phenomenon known as permeability and retention (EPR) effect (*52*). Tumor vessels have indeed bigger fenestration pores than normal vasculature, and tumors' lymphatic drainage network is inadequate to clear macromolecules that enter in the tumor area, thus both contributing to accumulation of systemically administered macromolecular compounds, including nanoparticles (*53*). New techniques are now being engineered to take into account this unique tumor microenvironment for imaging applications, although not necessarily at the single-cell level. For example, Merrimack Pharmaceuticals' MM-DX-929 is a nanoliposomal particle system functionalized with polyethylene glycol (PEG) and encapsulating copper-64 for positron emission tomography (PET) imaging function (*54*). MM-DX-929 is being developed as a "companion system" that helps identify patient populations who would best respond to liposomal chemotherapeutics (*55*).

While the examples highlighted above illustrate the progress made towards nanotechnology-enabled tumor diagnostics, it is important to note that the biological signature of cancer cells is highly heterogeneous and indeed varies from patient to patient (even within a cancer type). Identifying a correct combination of biomarkers for a particular cancer can be therefore particularly challenging. Future progress in nanotechnology-enabled cancer diagnostics should be synergized with advancements in genomics, protein engineering, and tumor biology.

NNI 2.0: Fueling the Engines of Creation

While investments in fundamental nanoscience and engineering will continue, the second era of the National Nanotechnology Initiative, dubbed NNI 2.0, will provide greater focus on realizing the full benefits of progress made in basic and applied research, including advances in each of the Clinton challenges highlighted above. According to the latest review of the NNI by the President's Council of Advisors on Science and Technology (PCAST), "this next technological generation will see the evolution from nanoscale components to interdisciplinary nano-systems and the movement from a foundational research-based initiative to one that also provides the necessary focus to ensure rapid commercialization of nanotechnology" (*56*). So far, over twenty billion dollars have been invested by the Federal agencies participating in the NNI towards fundamental and applied nanotechnology R&D; world-class characterization, testing, and fabrication facilities; education and workforce development; and efforts directed at understanding and controlling the environmental, health, and safety aspects of nanotechnology (*13*).

A Renewed Focus on Commercialization

Transfer of nanotechnology research into commercial applications requires advancements in manufacturing technologies that are scalable and cost-effective, as well as access to and retention of a skilled workforce (*13*). One of the four goals of the NNI is to "foster the transfer of new technologies into products for commercial and public benefit." A range of public and private initiatives to accelerate the commercialization of nanotechnology have produced new activities that address manufacturing challenges, promote the formation of new businesses or success of early stage businesses, and development of a skilled workforce (*57*). For example, a recent NNI-sponsored technical interchange meeting on Realizing the Promise of Carbon Nanotubes: Challenges, Opportunities, and the Pathway to Commercialization (*31*), focused on identifying the technical barriers to the production of CNT-based bulk and composite materials with electrical and mechanical properties nearer the ideal, and exploring ways to overcome these barriers.

Another example is the work by NNCO with the small- and medium-sized business community to coordinate a series of publicly-available webinars that identify challenges and successes in the commercialization of nanotechnology and provide information from the public and private sectors to help address these challenges (*58*). NNCO is also supporting Federal agencies participating in the NNI to launch multiple new activities aimed at educating students and the public about nanotechnology, including image and video contests highlighting student research (EnvisioNano), a new webinar series focused on providing nanotechnology information for K-12 teachers, and a searchable web portal of nanoscale science and engineering resources for teachers and professors on nano.gov.

All Nanotechnologies on Deck To Solve Grand Challenges

On June 17, 2015, the Office of Science and Technology Policy (OSTP) issued a request for information (RFI) (*59*) and an accompanying blog post (*60*) seeking suggestions for Nanotechnology-Inspired Grand Challenges for the Next Decade. As an element of President Obama's Strategy for American Innovation (*61*), a Grand Challenge is described as "an ambitious but achievable goal that requires advances in science and technology to achieve, and that has the potential to capture the public's imagination." Some historical examples of Grand Challenges have been President Kennedy's call to put a man on the moon or the Human Genome Project. Under the auspices of the NSET Subcommittee, Federal agencies participating in the NNI, working with OSTP and the NNCO, developed several examples of nanotechnology-inspired Grand Challenges. These examples were included in the RFI and covered broad topics in the areas of health care, electronics, materials, sustainability, and product safety.

An important component of these Grand Challenges is harnessing public–private mechanisms to achieve the commercialization of nanotechnologies. PCAST, in their most recent triennial assessment of the NNI, recommended "that the Federal Government transition its activities toward facilitating commercialization by directing the formulation of specific nanotechnology Grand Challenges. The Grand Challenges framework—a partnership between the public and private sectors—can drive scientific advances to revolutionary commercialized products" (*56*). A priority was also placed on ensuring that these Grand Challenges have a specific set of criteria, including: a well-defined goal inspiring different sectors and addressing an issue of significant societal impact; a measurable end-point and a finite lifetime; and a network of activities that will drive scientific ideas to commercialization and catalyze new discovery for technologies of the future. As representatives from OSTP and NNI member agencies review these submissions, a priority will be placed on establishing a process to engage with leaders from industry and private organizations to take on a Grand Challenge topic.

Conclusion

This chapter highlights several key aspects of the NNI, including the structure and history of the initiative, examples of progress towards achieving the Clinton challenges, and a brief review of future directions as the initiative enters its second era focused on commercialization. Interagency coordination and collaboration between Federal agencies participating in the NNI has resulted in shared knowledge and expertise within the initiative; joint engagement with stakeholders through sponsorship of solicitations and workshops; and leveraging of funding, staff, and facility resources. As expected, chemical sciences have played an important role in endowing researchers with exciting new tools and methods to understand and control matter at the nanoscale. The NNI has now matured into a partnership that is now home to twenty Federal department and agencies, and has resulted in remarkable progress in nanoscience as highlighted by examples of accomplishments towards the Clinton challenges. Federal agencies participating in the NNI now look to focus on the responsible translation of existing innovations into commercial products. Success will require building on current, as well as fostering new, partnerships and collaborations inside and outside the Federal government to ultimately pave the way for "a revolution in technology and industry that benefits society."

List of Acronyms

ARS	Agricultural Research Service
BIS	Bureau of Industry and Security
CDC	Centers for Disease Control and Prevention
CPSC	Consumer Product Safety Commission
CSREES	Cooperative State Research, Education, and Extension Service
DHHS	Department of Health and Human Services
DHS	Department of Homeland Security
DNI	Director of National Intelligence
DOC	Department of Commerce
DOD	Department of Defense
DOE	Department of Energy
DOEd	Department of Education
DOI	Department of the Interior
DOJ	Department of Justice
DOL	Department of Labor
DOS	Department of State
DOT	Department of Transportation
DOTreas	Department of the Treasury
EPA	Environmental Protection Agency
FDA	Food and Drug Administration
FHWA	Federal Highway Administration
FS	Forest Service
ITC	International Trade Commission
ITIC	Intelligence Technology Innovation Center
NASA	National Aeronautics and Space Administration
NEHI	Nanotechnology Environmental and Health Implications Working Group of the NSET Subcommittee
NICE	Nanotechnology Innovation and Commercialization Ecosystem Working Group of the NSET Subcommittee
NIFA	National Institute of Food and Agriculture
NIH	National Institutes of Health
NIOSH	National Institute for Occupational Safety and Health
NIST	National Institute of Standards and Technology
NNCO	National Nanotechnology Coordination Office
NNI	National Nanotechnology Initiative
NRC	Nuclear Regulatory Commission
NRO	National Reconnaissance Office

NSET	Nanoscale Science, Engineering, and Technology Subcommittee
NSF	National Science Foundation
NSTC	National Science and Technology Council
OMB	Office of Management and Budget
OSHA	Occupational Safety and Health Administration
OSTP	Office of Science and Technology Policy
PCAST	President's Council of Advisors on Science and Technology
TA	Technology Administration
USDA	U.S. Department of Agriculture
USGS	U.S. Geological Survey
USPTO	U.S. Patent and Trademark Office

References

1. Whitesides, G. M. *Small* **2005**, *1*, 172–179.
2. Chidsey, C. E. D. *Nanoscience: The Role of "Making" New Forms of Matter*; Stanford University: Stanford, CA,2006.
3. *What's So Special about the Nanoscale?* http://www.nano.gov/nanotech-101/special (accessed Oct 6, 2015).
4. Lane, N.; Kalil, T. The National Nanotechnology Initiative: Present at the Creation . *Issues in Science and Technology* **2005**, *21* (4).
5. Lewis, N. S. *Science* **2007**, *315*, 798–801.
6. Meza, L. R.; Das, S.; Greer, J. R. *Science* **2014**, *345*, 1322–1326.
7. McAlpine, M. C.; Ahmad, H.; Wang, D.; Heath, J. R. *Nat. Mater.* **2007**, *6*, 379–384.
8. Wright, J. *Nature* **2014**, *509*, S58–S59.
9. Heath, J. R.; Kuekes, P. J.; Snider, G. S.; Williams, R. S. *Science* **1998**, *280*, 1716–1721.
10. Nanoscale Science, Engineering, and Technology Subcommittee of the Committee on Technology. *National Nanotechnology Initiative Strategic Plan*; National Science and Technology Council: Washington, DC, 2014.
11. Clinton, W. J. *President Clinton's Address to Caltech on Science and Technology*; http://caltechcampuspubs.library.caltech.edu/2676/1/nano_clinton.pdf (accessed Oct 5, 2015).
12. Goddard, W. A.; Brenner, D.; Lyshevski, S. E.; Lafrate, G. J. *Handbook of Nanoscience, Engineering, and Technology*, 2nd ed.; CRC Press: Boca Raton, FL, 2007.
13. Nanoscale Science, Engineering, and Technology Subcommittee of the Committee on Technology. *The National Nanotechnology Initiative Supplement to the President's 2016 Budget*; National Science and Technology Council: Washington, DC, 2015.
14. *21st Century Nanotechnology Research and Development Act*; 2003; Vol. 117, pp 1923–1932.
15. *The Economist. Waiting for the Space Elevator*; http://www.economist.com/node/7001786 (accessed Oct 16, 2015).

16. Jones, R. *Nat. Nanotechnol.* **2007**, *2*, 453–454.
17. Meador, M., A. *Future Needs and Opportunities in Nanotechnology for Aerospace Applications: A NASA Perspective*; NIA Nanotechnology Workshop, Feb.21, 2014.
18. The Center for Nanoscale Science and Technology. *The New Steel? Enabling the Carbon Nanomaterials Revolution: Markets, Metrology, Safety, and Scale-up*; National Institute of Standards and Technology: Washington, DC, 2011.
19. Nanocomp Technologies, Inc. *Nanocomp Reaches New Frontiers on NASA's Juno Mission*; http://www.nanocomptech.com/nasa-juno-mission (accessed Oct 16, 2015).
20. Trimble, S. *Lockheed Martin reveals F-35 to feature nanocomposite structures* http://www.flightglobal.com/news/articles/lockheed-martin-reveals-f-35-to-feature-nanocomposite-357223/ (accessed Oct 16, 2014).
21. Vernon, D., M. *Enter The World Of Tomorrow: Nanotechnology*; http://www.metrocorpcounsel.com/articles/22557/enter-world-tomorrow-nanotechnology (accessed Oct 16, 2015).
22. Spitalsky, Z.; Tasis, D.; Papagelis, K.; Galiotis, C. *Prog. Polym. Sci.* **2010**, *35*, 357–401.
23. Zhang, M.; Atkinson, K. R.; Baughman, R. H. *Science* **2004**, *306*, 1358–1361.
24. Koziol, K.; Vilatela, J.; Moisala, A.; Motta, M.; Cunniff, P.; Sennett, M.; Windle, A. *Science* **2007**, *318*, 1892–1895.
25. Behabtu, N.; Young, C. C.; Tsentalovich, D. E.; Kleinerman, O.; Wang, X.; Ma, A. W. K.; Bengio, E. A.; Waarbeek, R. F.; Jong, J. J.; Hoogerwerf, R. E.; Fairchild, S. B.; Ferguson, J. B.; Maruyama, B.; Kono, J.; Talmon, Y.; Cohen, Y.; Otto, M. J.; Pasquali, M. *Science* **2013**, *339*, 182–186.
26. Jiang, C.; Saha, A.; Young, C. C.; Hashim, D. P.; Ramirez, C. E.; Ajayan, P. M.; Pasquali, M.; Martí, A. A. *ACS Nano* **2014**, *8*, 9107–9112.
27. Boncel, S.; Sundaram, R. M.; Windle, A. H.; Koziol, K. K. K. *ACS Nano* **2011**, *5*, 9339–9344.
28. Beese, A. M.; Wei, X.; Sarkar, S.; Ramachandramoorthy, R.; Roenbeck, M. R.; Moravsky, A.; Ford, M.; Yavari, F.; Keane, D. T.; Loutfy, R. O.; Nguyen, S. T.; Espinosa, H. D. *ACS Nano* **2014**, *8*, 11454–11466.
29. Miller, S. G.; Williams, T. S.; Baker, J. S.; Solá, F.; Lebron-Colon, M.; McCorkle, L. S.; Wilmoth, N. G.; Gaier, J.; Chen, M.; Meador, M. A. *ACS Appl. Mater. Interfaces* **2014**, *6*, 6120–6126.
30. Park, C.; Ounaies, Z.; Watson, K. A.; Pawlowski, K.; Lowther, S. E.; Connell, J. W.; Siochi, E. J.; Harrison, J. S.; Clair, T. L. S. In *Symposium Z – Making Functional Materials with Nanotubes*; MRS Online Proceedings Library; 2001; Vol. 706.
31. Nanoscale Science, Engineering, and Technology Subcommittee of the Committee on Technology. *Realizing the Promise of Carbon Nanotubes: Challenges, Opportunities, and the Pathway to Commercialization*; National Science and Technology Council: Washington, D.C., 2014.
32. *The Economist. All too much*; http://www.economist.com/node/15557421 (accessed Oct 6, 2015).

33. Markov, I. L. *Nature* **2014**, *512*, 147–154.
34. International Business Machines, Corp. *Atomic-Scale Magnetic Memory*; http://www.ibm.com/smarterplanet/global/files/us__en_us__computing__atomic_scale_magnetic_memory_011212.pdf (accessed Oct 6, 2015).
35. Zakeri, B.; Lu, T. K. *Curr. Opin. Chem. Biol.* **2015**, *28*, 9–14.
36. Zhang, F.; Nangreave, J.; Liu, Y.; Yan, H. *J. Am. Chem. Soc.* **2014**, *136*, 11198–11211.
37. Tripp, S.; Grueber, M. *Economic Impact of the Human Genome Project*; Battelle Memorial Institute: Columbus, OH, 2011.
38. Church, G. M.; Gao, Y.; Kosuri, S. *Science* **2012**, *337*, 1628–1628.
39. Lutz, J.-F.; Ouchi, M.; Liu, D. R.; Sawamoto, M. *Science* **2013**, *341*, 1238149.
40. Roy, R. K.; Meszynska, A.; Laure, C.; Charles, L.; Verchin, C.; Lutz, J.-F. *Nat. Commun.* **2015**, *6*, 7237.
41. Genzlinger, N. *Review: "In Cancer: The Emperor of All Maladies," Battling an Opportunistic Killer*; http://www.nytimes.com/2015/03/28/arts/television/review-in-cancer-the-emperor-of-all-maladies-battling-an-opportunistic-killer.html?_r=0 (accessed Oct 6, 2015).
42. American Cancer Society, Inc. *Cancer Facts and Figures 2015* http://www.cancer.org/acs/groups/content/@editorial/documents/document/acspc-044552.pdf (accessed Oct 6, 2015).
43. Schroeder, A.; Heller, D. A.; Winslow, M. M.; Dahlman, J. E.; Pratt, G. W.; Langer, R.; Jacks, T.; Anderson, D. G. *Nat. Rev. Cancer* **2012**, *12*, 39–50.
44. Plaks, V.; Koopman, C. D.; Werb, Z. *Science* **2013**, *341*.
45. VERIDEX, LLC. *510(k) Summary: CellSearchTM Circulating Tumor Cell Kit Premarket Notification-Expanded Indications for Use-Colorectal*; http://www.accessdata.fda.gov/cdrh_docs/pdf7/k071729.pdf (accessed Oct 6, 2015).
46. *About CELLSEARCH®: How does the CELLSEARCH® System Work?* https://www.cellsearchctc.com/about-cellsearch/how-cellsearch-ctc-test-works (accessed Oct 6, 2015).
47. Lowes, L. E.; Hedley, B. D.; Keeney, M.; Allan, A. L. *Cytometry A* **2012**, *81*, 983–995.
48. *Nanosphere. Verigene System*; http://www.nanosphere.us/products/verigene-system (accessed Oct 6, 2015).
49. Gibbons, W. *Protein Biomarker Capabilities for Nanosphere*, 2013.
50. Nanosphere. *Nanosphere to Release Positive Pilot Study Data for Ultra-Sensitive Protein Cancer Biomarkers in Conjunction With the American Society of Clinical Oncology (ASCO) 2011 Annual Meeting*; http://phx.corporate-ir.net/phoenix.zhtml?c=214748&p=irol-newsArticle&ID=1565592 (accessed Oct 6, 2015).
51. Metz, C. *Wired Magazine* **2014**.
52. Prabhakar, U.; Maeda, H.; Jain, R. K.; Sevick-Muraca, E. M.; Zamboni, W.; Farokhzad, O. C.; Barry, S. T.; Gabizon, A.; Grodzinski, P.; Blakey, D. C. *Cancer Res.* **2013**, *73*, 2412–2417.

53. Dimou, A.; Syrigos, K. N.; Saif, M. W. *Ther. Adv. Med. Oncol.* **2012**, *4*, 271–279.
54. Lee, H. *A Novel ^{64}Cu-Liposomal PET Agent (MM-DX-929) Predicts Response to Liposomal Chemotherapeutics in Preclinical Breast Cancer Models*, 2012.
55. Merrimack. *Merrimack Pharmaceuticals Presents Promising Phase I Clinical Data for Novel Agent to Treat HER2-positive Breast Cancer and Preclinical Data on Companion Diagnostic*; http://investors.merrimackpharma.com/releasedetail.cfm?ReleaseID=725828 (accessed Oct 6, 2015).
56. President's Council of Advisors on Science and Technology. *Report to the President and Congress on the Fifth Assessment of the National Nanotechnology Initiative*; Executive Office of the President: Washington, DC, 2014.
57. Whitman, L. *New Initiatives to Accelerate the Commercialization of Nanotechnology*; http://www.nano.gov/May2015Forum (accessed Oct 6, 2015).
58. *NNCO Small- and Medium-Sized Enterprise (SME) Webinar Series* http://www.nano.gov/SMEwebinars2015 (accessed Oct 6, 2015).
59. Office of Science and Technology Policy. *Federal Register: Request for Information: Nanotechnology-Inspired Grand Challenges for the Next Decade*; https://federalregister.gov/a/2015-14914 (accessed Nov 4, 2014).
60. Whitman, L.; Kalil, T. *A Call for Nanotechnology-Inspired Grand Challenges*; https://www.whitehouse.gov/blog/2015/06/17/call-nanotechnology-inspired-grand-challenges (accessed Oct 6, 2015).
61. *A Strategy for American Innovation: Securing Our Economic Growth and Prosperity*; https://www.whitehouse.gov/node/25867 (accessed Oct 6, 2015).

Chapter 4

Building Foundational Knowledge and Infrastructure for Nanotechnology: 2000–2030

Mihail C. Roco*

National Science Foundation & National Nanotechnology Initiative, 4201 Wilson Boulevard, Arlington, Virginia 22230, United States
***E-mail: mroco@nsf.gov.**

The progression of nanoscale science and technology development may be separated in three main stages: Nano 1 (~2000-2010) focused on uncovering phenomena at the nanoscale and semi-empirical synthesis of nanocomponents; Nano 2 (~2010-2020) on science-based nanosystem integration for fundamentally new products; and Nano 3 (~2020-2030) on creating new nanosystem architectures and converging technology platforms. This paper outlines how the foundational nanoscale science and engineering field has advanced in basic knowledge, infrastructure and numerous economic sectors, such as catalysts, semiconductor manufacturing, medicine, agriculture, and energy production. Global revenues of nano-enabled products reached $1 trillion in 2013 according to industry surveys. The annual U.S. and global revenue growth rate increased from about 25% during 2001-2010 to about 40% during 2010-2014, suggesting that this recent interval is the beginning of the ascendant section of the S-development curve.

Introduction

Nanotechnology is the ability to control and restructure matter at the atomic and molecular levels in the range of approximately 1–100 nm, and to exploit distinct nanoscale properties and phenomena. The aim is to create materials, devices, and systems with fundamentally new properties and functions for novel applications by engineering their small structure. The unifying definition of nanotechnology, based on control of matter at the nanoscale, and the long-term vision for nanotechnology research, education and innovation, were formulated during 1997-1999 and began to be implemented with the National Nanotechnology Initiative (NNI) in 2000. According to estimates, nanotechnology would take about three decades to advance from a scientific curiosity in 2000 to a science-based general purpose technology with broad societal benefits in around 2030 (*1–3*).

A long-term strategic view is needed for the development of a foundational general purpose S&T field. For nanotechnology, the vision is unfolding in three developmental stages that correspond with the complexity of typical processes, tools, and outcomes: passive and active nanostructures in the first stage of development (Nano 1), nanosystems and molecular nanosystems in the second stage (Nano 2), and converging technology platforms and distributed interconnected nanosystems in last stage (Nano 3).

This paper outlines the long-view of establishing knowledge and infrastructure for nanotechnology between 2000 and 2030.

Nanotechnology – A Foundational Megatrend in Science and Engineering

Nanotechnology is a foundational, general purpose S&T field for all sectors of the economy dealing with matter and biosystems, as information technology is a general-purpose technology for communication and computation. Two other foundational technologies emerging at the beginning of the 21st century are biotechnology and cognitive technologies (Figure 1). These are the four foundational megatrends in science and engineering in the first quarter of the 21st century. New specific S&T fields are created continuously at the confluence, by spin-off and recombination of the four foundational NBIC (nano-bio-info-cognitive) fields.

Nanotechnology continues its quasi-exponential growth: by vertical science-to-technology-to-production transition in advanced materials, pharmaceuticals, and electronics; horizontal expansion to areas such as agriculture, textiles, and cement; and many spin-off areas (~20) such as spintronics, metamaterials, and plasmonics. Nanotechnology is progressively penetrating diverse industries and economic sectors. For this reason, it is increasingly difficult to identify the R&D programs around the word that support nanotechnology, because their names reflect an activity or field that has branched out of the foundational field.

Overall government expenditures for nanotechnology including the branched out fields reached about $9.5 billion in 2012 (Figure 2). The main global characteristic of these expenditures is rapid and uneven growth among countries and groups of countries, with relatively faster growth for the BRIC countries (Brazil, Russia, India, and China).

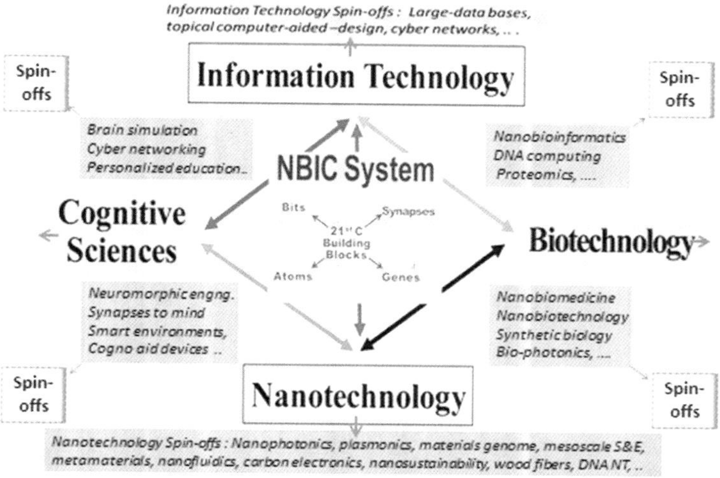

Figure 1. Converging foundational technologies and their inter-disciplinary and spin-off S&T fields. Adapted with permission from ref. (4). Copyright 2013, Springer.)

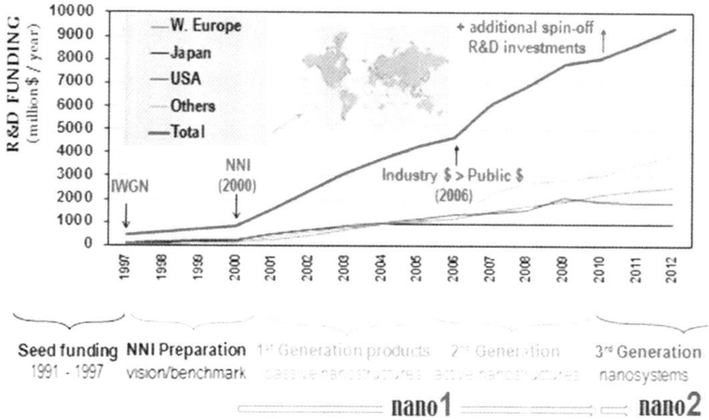

Figure 2. International government R&D funding for interval 2000- 2012 (using NNI definition, from 81 countries, collected by direct contacts); after 2013 – the increase use of new terms such as spintronics and metamaterials instead of "nano" makes such searches less precise.*

In the United States and abroad, most of the larger science and technology initiatives have been justified mainly by application-related and societal factors. Leading examples are the Manhattan Project during World War II (with centralized, goal focused, simultaneous approaches), the Apollo Space Project (with a centralized, focused goal), and Networking and Information Technology Research and Development (top-down initiated and managed, and established when mass applications justified the return of investment). The initiation of the National Nanotechnology Initiative (NNI) was motivated primarily by its long-term science and engineering goals and the opportunity for general-purpose technology, and NNI has been managed using a bottom-up approach combined with centralized coordination. In the words of Charles Vest, President National Academy of Engineering (PCAST meeting, White House, 2005), "NNI is a new way to run an initiative."

Nanotechnology promises to become a general purpose technology with large scale applications similar to digital technology. It could eventually match or outstrip the digital revolution in terms of economic importance and societal impact once the methods of investigation and manufacturing are developed and the underlying education programs and physical infrastructure are established. During about 2020-2030, nanotechnology could equal and even exceed the digital revolution in terms of technology breakthroughs, level of investments, and societal importance (Figure 3) (*5*).

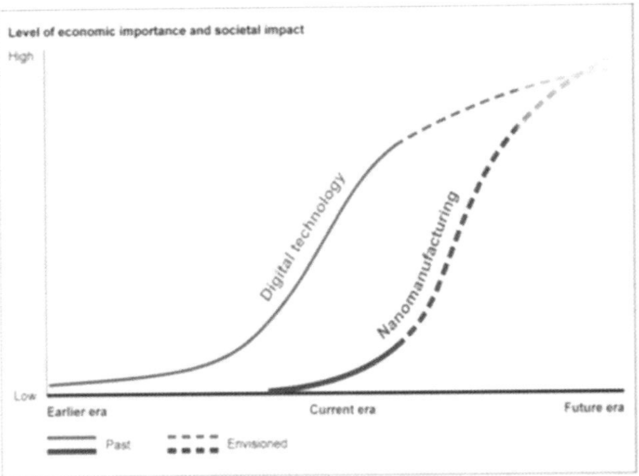

Figure 3. S-curves for two science and technology megatrends: Past (solid lines until 2014) and envisioned (dashed lines) conceptualization of "Nanomanufacturing" and "Digital Technology". (5)

The nanotechnology development S-curve shown in Figure 3 is supported by the data on revenues from products incorporating nanotechnology showing an increase of the world annual rate of revenues growth from 25% in 2000-2010 (*2*) to over 40% in 2010-2013 (*6*).

The President Council of Advisors in Science and Technology (PCAST) biennial report (2014) made several recommendations to support a 15-year vision for NNI through 2030 (*7*), including:

- A healthy research environment and infrastructure have been leading to transformative discovery and must continue to be nurtured;
- The nanotechnology community is transitioning and needs a strong ecosystem to ensure competitive technologies reach the commercial marketplace; and
- The primary active structure for R&D program management should be driven by the Federal and OSTP nanotechnology Grand Challenges.

Following these recommendations, NNI has developed a process for identification of nano-enabled grand challenges suitable for the second stage of nanotechnology development (Nano 2), which is focused on nanosystems and societal implications.

Nanotechnology-enabled products by sectors with most traded nano-products in 2014 are (a) nanostructured materials and manufacturing (including production and use of catalysts, coatings, fiber/nanotube reinforces plastics, imprinting, and roll-to-roll manufacturing); (b) nano-electronics, magnetics and photonics (including semiconductors, mobile electronics and displays, packaging, thermal management, batteries, supercapacitors), (c) healthcare applications (including diagnostic and monitoring sensors, cosmetics, food products and packaging, personal care products, sunscreen, packaging, surgical tools, implantable medical devices, contrast agents, and drug delivery systems); and (d) energy and environment (including improvements in oil industry and environmental clean-up, fuel cells, solar cells, filtration, sensors, and water treatment and purification.)

Three Stages for Establishing the New General Purpose Technology

How long is needed for nanotechnology to become a general purpose technology — where the methods, tools, manufacturing chain, physical and education infrastructure are readily available for general use? We have estimated that about 30 years are needed for nanotechnology development to evolve from a scientific curiosity to general use in the economy. This period may be separated in to three stages that correspond to the level of complexity and typical outcomes. Each stage is defined by its investigative methods; techniques for synthesis/assembling/manufacturing; level of nanoscale integration and complexity of the respective products; typical application areas; education needs; and risk governance. These characteristics are documented in four reports shown in Figure 4 (*1–3*, *8*). The reports are available on: www.wtec.org/nano2/ and www.wtec.org/NBIC2-report/. Outcomes halfway through the cited development interval generally confirm the long-term vision (*9*).

Figure 4. Thirty year vision to establish nanotechnology: changing the research and education focus and priorities in three stages from scientific curiosity to immersion in socioeconomic projects. Adapted with permission from ref. (4). Copyright 2013, Springer.)

A schematic of the three nanotechnology development stages and corresponding generations of products incorporating nanostructures are shown in Figure 5.

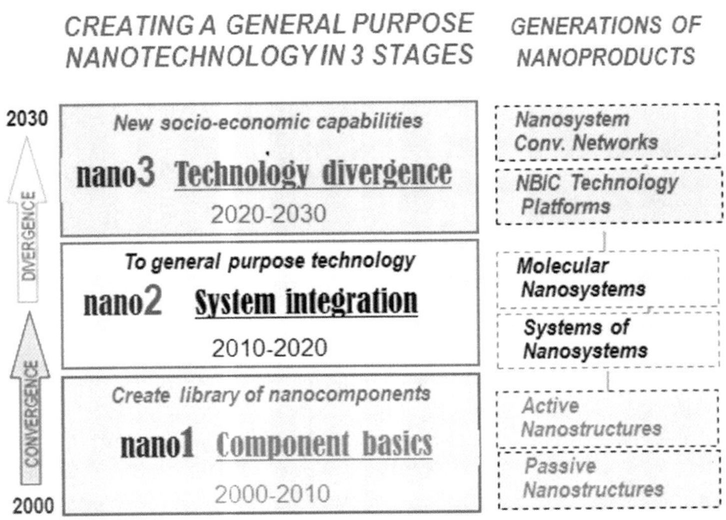

Figure 5. Creating a general purpose technology in three development stages between 2000 and 2030. Each stage includes introduction of two generations of nanotechnology products.

Nano 1: Component Basics

The planning in the first stage of nanotechnology development (Nano 1) was focused on discovery of individual phenomena and semi-empirical synthesis of nanoscale components. The main defining phenomena at the nanoscale were identified and at least partially quantified. For example, quantum effect repulsion forces between surfaces at small interfaces were identified, and the first quantum device was built. In another example, quantum dots, nanoparticles, nanotubes and nanocoatings were created from a majority of elements in the periodic table, and their atomic and molecular assembling mechanisms were explored for generalizations. Typical measurements of the atomic structure with femtosecond changes were still indirect, using time and volume averaging. The nanocomponents typically were used to improve performance of existing products, such as nanoparticle reinforced polymers. The main technology sector penetration of nanotechnology was in advanced materials, nanostructured chemicals, electronics and pharmaceutics. For example, Moore's Law for semiconductors has continued because of nanoscale components added in their fabrication.

Education transitioned from focus on microscale foundation where the properties of materials and devices are fixed to nanoscale-based understanding of nature and technology where properties may be changes and engineered as a function of matter arrangement at the nanoscale. Governance was mostly science-centric with a focus on nanotechnology environmental, health and safety aspects. National and international organizations formulated the basics for nomenclature to be used in scientific publications and for standards classifications. An international, multidisciplinary nanoscale science and technology community became established.

Various generations of nanotechnology products and productive processes are timed with the introduction of new prototypes of nanotechnology products and with the successive increases in the degree of control, integration, complexity and risk. In the first stage of nanotechnology development (2000-2010) one can identify two generations of products and productive processes:

Generation 1. ***Passive Nanostructures*** (~2000-2005): The nanostructures have stable behavior during their use. They typically are used to tailor macroscale properties and functions.

a. Dispersed nanostructures, such as nanoscale aerosols and colloids, quantum dots on surfaces, various dispersed nanotubes and monofilaments
b. Contact nanostructures, such as in nano-composites, metals, polymers, ceramics, and coatings

Generation 2. ***Active Nanostructures*** (~2005-2010): The nanostructures change their composition and/or behavior during their use. They typically are integrated into microscale devices and systems and used for their biological, mechanical, electronic, magnetic, photonic, and other effects.

a. Bio-active nanostructures with health effects, such as targeted drugs, biodevices, and artificial muscles
b. Physical-chemical active nanostructures, such as amplifiers, actuators, state-changing sensors, adaptive structures, 3D transistors

Table 1. Main NSF R&D Centers, Networks, and User Facilities for NNI

Name	*Institution(s)*
NSF Ten Networks	
National Nanofabrication Infrastructure Network (NNIN) – 15 nodes (user facilities) (www.nnin.org) National Nanotechnology Coordinated Infrastructure (NNCI) after Sept. 2015 – 16 nodes	Cornell University – main node (NNIN re-competed in 2015)
Network for Computational Nanotechnology (NCN) (nanoHUB.org)	Purdue University – main node
National Nanomanufacturing Network (NNN) (www.internano.org)	University of Massachusetts, Amherst – main node
Centers for Nanotechnology in Society (CNS) (cns.asu.edu)	Arizona State University and University of California, San Diego
Nanoscale Informal Science Education (NISE) Network (www.nisenet.org)	Museum of Science, Boston – main node
Nanoscale Science and Engineering Centers (NSEC)	Distributed centers
Materials Science and Engineering Centers (MRSECs)	Distributed centers across U.S.
Nanosystems Engineering Research Centers (NERC)	Distributed centers across U.S.
Centers for the Environmental Implications of Nanotechnology (CEIN) (www.cein.ucla.edu)	University of California, Los Angeles, and Duke University
Center for National Nanotechnology Applications and Career Knowledge (NACK) (nano4me.org)	Pennsylvania State University
NSF Three Science and Technology Centers	
Center for Energy Efficient Electronics Science (nanoelectronics) (www.e3s-center.org)	University of California, Berkeley
Emergent Behaviors of Integrated Cellular Systems (nanobiotechnology) (http://ebics.net/about)	Massachusetts Institute of Technology
Center for Integrated Quantum Materials (ciqm.harvard.edu)	Harvard University

An important result of the investment in nanotechnology has been the creation of a flexible infrastructure. PCAST (2010) (*10*) report notes that in the interval 2001-2010 "The NNI has created outstanding resources for the nanotechnology R&D community through its investments in infrastructure such as shared user facilities, research centers and networks, and education centers and networks. For illustration, NSF has created during the first stage of nanotechnology development (Nano 1) ten R&D centers, networks and user facilities (Table 1).

Key infrastructure centers, networks and facilities established by DOE, NIH, DOD and NIST are shown in Tables 2 to 5.

Table 2. DOE Network of Five NNI User Facilities

Name	*Institution(s)*
Center for Functional Nanomaterials (www.bnl.gov/cfn)	Brookhaven National Laboratory
Center for Integrated Nanotechnologies (cint.lanl.gov)	Sandia National Laboratory and Los Alamos National Laboratory
Center for Nanophase Materials Sciences (cnms.ornl.gov)	Oak Ridge National Laboratory
Center for Nanoscale Materials (www.anl.gov/cnm)	Argonne National Laboratory
Center for Molecular Foundry (foundry.lbl.gov)	Lawrence Berkeley National Laboratory

Table 3. NIH Network of Four NNI Centers and Networks

Name	*Institution(s)*
NHLBI Centers of Excellence in Nanotechnology (grants.nih.gov/grants/guide/rfa-files/RFA-HL-04-020.html)	Four distributed centers
Nanomedicine Development Centers (grants.nih.gov/grants/guide/rfa-files/RFA-RM-06-007.html)	Eight distributed centers
Centers of Cancer Nanotechnology Excellence (www.cancercenter.com)	Eight distributed centers
Nanotechnology Characterization Laboratory (user facilities) (ncl.cancer.gov)	Frederick, Maryland campus

Table 4. DOD NNI Institutes, Facilities, and Consortia

Name	*Institution(s)*
Naval Research Laboratory Institute for Nanoscience	Internal user facility and an important part of NRL's infrastructure
The Institute for Soldier Nanotechnologies at MIT	Eight distributed centers
Army Nanopowder Pilot Production Facility	At the Armament Research, Development, and Engineering Center, Picatinny Arsenal
Nano-Bio Manufacturing Consortium (NBMC)	common platform to address flexible device applications; AFRL with 50% industry cost share; new ecosystem

Table 5. NIST User Facilities in Support of NNI

Name	*Institution(s)*
Center for Nanoscale Science and Technology (CNST), shared-use nanofabrication facility (NanoFab) (www.nist.gov/cnst)	Gaithersburg, Maryland campus
NIST Center for Neutron Research (NCNR)	Gaithersburg, Maryland campus
NIST Boulder Precision Measurement Laboratory (partial nano)	Boulder campus
National Synchrotron Light Source-II at Brookhaven National Laboratory	NIST-DOE partnership (at DOE laboratory)

Nano 2: System Integration

During the second stage of nanotechnology development (Nano 2), the focus is on integration at the nanoscale and science-based creation of devices and systems for fundamentally new products, including self-powered nanodevices, hierarchical self-assembling systems, and nano-bio assemblies. Examples include nanofluidics systems, integrated sensorial systems, and nanoelectronics and display systems. Direct measurements and simulations with atomic and femtosecond resolutions have been undertaken for many-atom systems encountered in biological and engineering applications. There is an increased focus on new performance in new domains of application, and on innovation methods. Nanotechnology penetration is faster in nanobiotechnology, energy resources, food and agriculture, forestry, and cognitive technologies, as well as in simulation-based design methods.

In education we see more attention to cross-disciplinary "T" or "reverse T" learning, including general nanotechnology education in the horizontal component. Societal aspects are increasingly on expanding sustainability, exploiting the potential for increased productivity and addressing socio-economic issues with a focus on healthcare. Governance is increasingly user-centric and multiple player participatory. Global implications are seen in economics, sustainability and international balance of forces.

In the second stage of nanotechnology development (2010-2020), one also can identify two generations of nano products and productive processes:

Generation 3. ***System of Nanosystems*** (~2010-2015): The three-dimensional nanosystems will use various synthesis and assembling techniques such as bio-assembling, synthetic chemistry, nanomodular and hybrid assembling, robotics with emerging behavior, and evolutionary approaches. They will be incorporated in to other larger hierarchical systems suitable to various technologies and application areas. Key challenges are networking of components at the nanoscale and hierarchical architectures. Research focus will shift towards heterogeneous nanostructures and supramolecular system engineering. This includes directed multiscale selfassembling, artificial tissues and sensorial systems, quantum interactions within nanoscale systems, processing of information using photons or electron spin, assemblies of nanoscale electromechanical systems (NEMS).

Generation 4. ***Molecular Nanosystems*** (~2015-2020): This will include products with heterogeneous molecular nanosystems where each molecule in the nanosystem has a specific structure and plays a different role. Molecules will be used as devices, and fundamentally new functions will emerge from their engineered structures and architectures. Designing new atomic and molecular assemblies is expected to increase in importance, including macromolecules "by design," nanoscale machines, and directed and multiscale self-assembling, exploiting quantum control, nanosystem biology for healthcare, modular nanotechnology, and human-machine interface at the tissue and nervous system level. Research will include topics such as: atomic manipulation for design of molecules and supramolecular systems, controlled interaction between light and matter with relevance to energy conversion among others, exploiting quantum control mechanical-chemical molecular processes, nanosystem biology for healthcare and agricultural systems, and human-machine interface at the tissue and nervous system level.

Priorities of the U.S. National Nanotechnology Initiative at the beginning of *Nano 2* (*2*) were defined under several "Signature Initiatives" since 2011: (a) Nanoelectronics for 2020 and Beyond; (b) Sustainable Nanomanufacturing; (c) Nanotechnology for Solar Energy; (d) Nanotechnology Knowledge Infrastructure, and (e) Nanosensors. PCAST (2014) (*7*) recommended the establishment of a new set of grand challenges for Nano 3 looking 15 years ahead to 2030.

A focus in 2015 is research on the third generation of nanotechnology products including nanosystems, self-powered nanodevices, and nano-bio assemblies. There is an increased focus on nanoscale science and engineering integration with other knowledge and technology domains to create new nanosystem architectures and corresponding technology platforms by 2030 ("Converging knowledge, technology and Society: Beyond Nano-Bio-Info-Cognitive Technologies", Springer 2013, available on www.wtec.org/NBIC2/).

New kinds of infrastructure are growing in importance during Nano 2: for sustainable nanomanufacturing including digital and biology components (*7*, *11*), modeling and simulation (e.g., materials genome, (see NSTC 2011 (*12*), and nanotechnology knowledge and informatics (see http://www.nano.gov/node/825).

Nano 3: Technology Divergence

After 2010, there is an increased focus on nanoscale science and engineering integration with other knowledge and technology domains and their applications (*4*) that will continue through the end of the third decade of nanotechnology development (Nano 3). After about 2000, convergence of nanotechnology with other key technologies has begun leading to bifurcation into emerging and integrated technology platforms. NBIC-based measurements will be needed in these new technology platforms. Integration of foundational and general technologies will branch out to new fields of research and production.

Education will need to be more focused on unifying concepts and connecting phenomena, processes and technologies. The role of bottom-up and horizontal interactions will increase in importance as compared to top down measures in the S&T governance. The risk analysis will expand to hybrid bio-nano systems and human-technology coevolution. New competencies, socio-economic platforms, and production capabilities will be take a significant role in the economy. The international community will be more connected through the scientific and technological developments. It will create opportunities for new models of collaboration and competition.

Two generations of nanoproducts and productive processes are envisioned in the third stage of nanotechnology development (2020-2030):

Generation 5. ***NBIC Integrated Technology Platforms*** (~2020-2025): Converging technology platforms from the nanoscale based on new nanosystem architectures will be established at confluence with other foundational emerging technologies. This includes converging foundational technologies (nano-bio-info-cogno) platforms integrated from the nanoscale.

Generation 6. ***Nanosystem Convergence Networks*** (~2025-2030): Distributed and interconnected nanosystem networks, across domains and interacting at various levels (foundational, topical, application, products/service), will be created for health, production, infrastructure and services. This includes networks of foundational technologies (nano-bio-info-cogno) platforms and their spin-offs including for emerging nano-biosystems.

A dominant trend in the interval 2020-2030 is envisioned to be immersion of nanotechnology with other emerging and established technologies, in industry, medicine and services, and in education and training for societal progress to become the largest technology driver in most economical sectors together with information technology.

Conclusions

Nanotechnology development has become an international scientific and technological endeavor with focused R&D programs in over 80 countries after the announcement of the National Nanotechnology Initiative in the United States in 2000. The long-term vision and collaboration among the national and international programs are essential factors in this global development. Among the major developments in science and technology at present, nanotechnology positions itself as the most exploratory of them. In nanoscale science and engineering there are emerging methodologies, supporting infrastructure, and education to create useful materials, devices and nanosystems – giving it perhaps the greatest scope for discovery, manipulation and diversification in coming years.

Nanotechnology is still in the formation phase of creating nanosystems by design for fundamentally new products. As we are at the beginning of the S-curve of nanotechnology development (GAO 2014), the main challenge is to prepare the knowledge base, manufacturing, people, physical infrastructure and anticipatory governance for the nanotechnology of tomorrow. This phase of nanotechnology development needs investment in dedicated R&D programs on new nanotechnology methods and system architectures suitable for various economy sectors, adaptation of education programs and physical infrastructure to its unifying and integrative concepts, and institutionalization of nanotechnology in relevant societal institutions. Global nanotechnology-based labor and markets are estimated to double approximately every three years. Nanotechnology will reach economic, large-scale applications, including convergence with other foundational technologies and their spin-offs domains, by 2030, reaching a level of development and use comparable to what information technology and the Internet achieved by about 2000.

Acknowledgments

This chapter is based on the author's experience in the nanotechnology field, as founding chair of the NSET subcommittee coordinating the NNI and as a result of interactions in international nanotechnology policy arenas. The opinions expressed here are those of the author and do not necessarily reflect the position of the U.S. National Science and Technology Council (NSTC) Subcommittee on Nanoscale Science, Engineering and Technology (NSET) or the National Science Foundation (NSF).

References

1. *Nanotechnology Research Directions: Vision for the Next Decade*; Roco, M. C., Williams, R. S., Alivisatos, A. P., Eds.; IWGN Workshop Report 1999, National Science and Technology Council, Washington, DC; Springer: Dordrecht, 2000; http://www.wtec.org/loyola/nano/IWGN.Research.Directions/.
2. Roco, M. C.; Mirkin, C.; Hersam, M. C. Nanotechnology Research Directions for Societal Needs in 2020. *J. Nanoparticle Res.* **2011**, *13*

(3), 897–919 (full report with the same title was published by Springer, Dordrecht, 2011, 670p).

3. Roco, M. C.; Harthorn, B.; Guston, D.; Shapira, P. Innovative and Responsible Governance of Nanotechnology for Societal Development. In *Nanotechnology Research Directions for Societal Needs in 2020*; Springer: Dordrecht, 2011; pp 561–618, available on http://www.cein.ucla.edu/new/file_uploads/chapter13.pdf.
4. Roco, M. C.; Bainbridge, W. S.; Tonn, B.; Whitesides, G. *Converging Knowledge, Technology and Society: Beyond Nano-Bio-Info-Cognitive Technologies*; Springer: New York, 2013, available on www.wtec.org/NBIC2-Report/.
5. GAO (Government Accountability Office). *Forum on Nanomanufacturing*; Report to Congress, GAO-14-181SP, 2014.
6. Lux Research. *Nanotechnology Update: Corporations up Their Spending as Revenues for Nano-enabled Products Increase*; Flynn, H., Hwang, D., Holman, M., Eds.; 2014; 18 p.
7. PCAST. *Report to the President and Congress on the Fifth Assessment of NNI*; White House: Washington, DC, October 2014; http://www.whitehouse.gov/sites/default/files/microsites/ostp/PCAST/pcast_fifth_nni_review_oct2014_final.pdf.
8. Roco, M. C.; Bainbridge S. M. *Converging Technologies for Improving Human Performance: Nanotechnology, Biotechnology, Information Technology and Cognitive Sciences*; Springer: New York, 2003.
9. Roco, M. C. *Affirmation of nanotechnology in economy, Frontiers of nanotechnology*; Mensah, T., Ed.; AIChE-Wiley: New York, 2015.
10. PCAST. *President's Council of Advisors on Science and Technology (PCAST), Third Assessment along with Recommendations of the National Nanotechnology Advisory Panel*; 2010; www.nano.gov/node/623.
11. PCAST. *President's Council of Advisors on Science and Technology (PCAST), Ensuring American Leadership in Advanced Manufacturing (2012)*; 2012; www.whitehouse.gov/administration/eop/ostp/pcast/docsreports.
12. National Science and Technology Council. *Materials Genome Initiative*; https://www.whitehouse.gov/sites/default/files/microsites/ostp/materials_genome_initiative-final.pdf.

Chapter 5

Nanomaterials and Nanomanufacturing with an Emphasis on National Security

Lewis E. Sloter*

Office of the Assistant Secretary of Defense for Research and Engineering, Department of Defense, Alexandria, Virginia 22350-3600, United States
***E-mail: lewis.e.sloter.civ@mail.mil.**

The status and ongoing role of nanomaterials and manufacturing research and evolving applications are presented in the context of the Department of Defense's (DoD) national security mission. Materials and processes highlights associated with the Military Departments and Defense Agencies are provided with a special emphasis on the breadth of opportunities and potentially enhanced or new defense capabilities and broader societal value provided by advances in nanotechnology generally and engineered nanomaterials specifically. The role of the DoD as a founding participant in the National Nanotechnology Initiative and the value of federal coordination and collaboration (as well as broader forms of collaboration, including international) are discussed along with areas of increasing emphasis, such as nanomanufacturing and the incorporation of nanomaterials in complex materials systems for functional advantage. The importance of a wide spectrum of participants, including academia, small businesses, large businesses, and consortia is noted in conjunction with selected examples.

DoD Research and Engineering (R&E)

The Department of Defense's mission is the maintenance and protection of national security. Research is conducted to support that mission. In general, the DoD strategy has "Three Enduring Principles": mitigate or eliminate current and emerging threats to national security, affordably enable new or extended military capabilities, and create technology surprise through science and engineering (*1*).

Several years ago, the DoD examined several different areas as part of that strategy: synthetic biology, modeling human behavior, engineered materials, cognitive neuroscience, quantum information and control, and nanoscience and engineering. At least three of these items are materials-intensive, and one is clearly focused on nano science and technology.

The DoD nanotech activities do not constitute a program in the sense that they allocate funding in advance for nanotechnology and science efforts. Those projects compete with all other science and technology within the DoD's research and technology budget. It is based on scientific merit -- quality science provides deep understanding and establishes the foundation for future research products and applications. The projects must also have relevance to the DoD to discover and exploit unique phenomena at these dimensions to enable novel applications enhancing soldier and weapon systems capabilities.

The DoD views nanotechnology as an important tool for developing future defense products. Nanotechnology investment composes about 45% of the DoD's fundamental research. The DoD does not appropriate money for nanotechnology, specifically. It views nanotechnology as an enabling technology that should receive the highest level of corporate attention and coordination.

Science and technology (S&T) programs do support projects in nanotechnology. Some of the really big ones are the Multidisciplinary University Research Initiative, which gave out seven nanoscale awards in FY14; the National Security Science and Engineering Facility Fellowship Program, which provided two nanoscience awards in FY14; All Military DoD agency S&T programs; the Small Business Innovation Research (SBIR) Program, which had an FY13 investment of about $22.5 million; the Small Business and Technology Transfer Program (STTR), which had an FY13 investment of about $5.5 million; and Manufacturing Technology Programs (*2*).

The DoD maintains two institutes that are directly associated with nanotechnology. One such institute is at the Naval Research Laboratory (*3*). Its mission is to conduct highly innovative, interdisciplinary research at the intersections of the fields of materials, electronics and biology in the nanometer size domain. Their research focuses on high-strength, low-weight materials; high-speed, low-power electronics; molecular sensors; and energy storage and conversion.

The other institute is the Army-MIT Institute for Soldier Nanotechologies. The university-affiliated research center invests in soldier protection, receives industry partnerships and participation, and accelerates the transition of research products. Their goal is to enhance objective force warrior survivability and leverage breakthroughs in nanoscience and nanomanufacturing. They have invested in nanofibers for lighter materials, active/reactive ballistic protection, environmental protection, directed energy protection, microclimate conditioning, signature management, chem/bio detection and protection, biomonitoring/triage, exoskeleton components, and forward counter mines (*4*).

Nanotechnology is associated and critically dependent on the processes used and the ability to control structure at the nanoscale. The army has put together a long-range, strategic document and approach to the utilization of nanotechnology.

It follows fundamental research towards applied research to prototypical products and finally to applications (Table 1).

A Nanomaterials Grand Challenge roadmap (Figure 1) provides a long-term view of the possible growth and development of nanomaterials, many of them possibly with relevance to DoD.

Associated Projects and Programs

One associated program is developing boron carbide nanofiber application. The approach uses boron carbide particulates at the nanoscale, a cellulosic matrix and traditional fiber-spinning processing, prior to pyrolysis, to develop nanograined boron carbide fiber. This provides the DoD the opportunity to incorporate this material into structures for blast resistance, penetration resistance, and for strength and stiffness associated with a high-modulus fiber.

Another example is environmental research coupled with highly defense-specific nanoenergetic materials for propulsion and explosive systems. Nanoparticulate aluminum and nanothermites are used in these applications. This particular project examines the toxicity of nanothermite residue from firing, collected within the cloud of fugitive particulates, and the results provide guidance to Army's decision on the development of nanotermites.

The area of nanotechnology could revolutionize transparent materials. Another project is nanocomposite optical ceramics, a material comparable to, or better than, sapphire which is the standard, medium-range infrared (MWIR) material. Sapphire, however, has high emissivity at higher temperatures and very poor machinability. The aluminum-free nanostructured oxide composites that have been developed are MWIR transparent, nearly as hard and strong as sapphire, easily processed and polished, and exhibit at least an order of magnitude lower emissivity. Nanocomposite ceramic domes are scheduled for transition to existing DoD systems.

There is a lot of potential in templating for nanomaterials' growth. One project has capitalized on this, creating a carbon nanotube (CNT) with a "near-perfect" black body by throwing an array of directional CNTs on substrates. Potential applications include infrared thermometers, light absorbers, optical microscopes, and camera mirrors.

Nanohybrid materials for radiation detection have also been developed through a DoD program. It attempts to create a much more affordable gamma detector. Currently, semiconductor detectors or solid-state hosts for scintillating materials are used mostly for gamma detection. The new, nanohybrid materials provide a relatively low-cost, flexible, and integral nanodetector.

Nano-Bio manufacturing consortium (NBMC) brings together a consortium in the nanotech and the confluence of biotech areas. It is a complex public-private partnership creating a valuable instrument for the next phases of nanotechnology integration and convergence. Examples of projects include biomarker sensors, biometric sensors, and system packaging.

Table 1. Examples of Nanotechnology of Interest to DoD

Examples of Potential DoD Applications of Nanotechnology

- Electronics and Sensing
  - Infrared focal plane arrays
- Power and Energy
  - Fuel-cell catalysts
- Structural Materials
  - "Fuzzy" carbon fibers
- Coatings
  - Photoactive, self-cleaning films
- Multifunctional Materials and Devices
  - Spin-polarized active devices
- Materials and Systems Prognosis
  - Quantum-dot thermography
- Energetics
  - Nano-sized aluminum powders, nanostructured reactive materials
- Chemical/Biological Defense
  - Chemical sensors based on non-Boltzmann heat flow

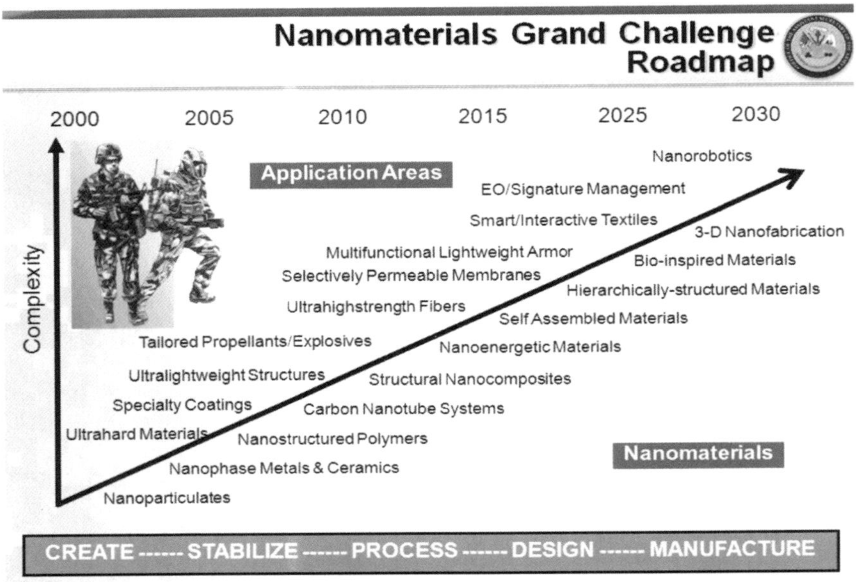

Figure 1. Nanomaterials Grand Challenge roadmap.

Another project attempts to move from 2D to 3D by printing 3D CNT structures on graphene. This provides a continuous semiconductor with opportunities for structural hybrids with controlled electronic properties.

A number of researchers are examining the potential of nanoparticulate and nanofiber reinforcements in the resin in carbon fiber resin matrix composites. The Air Force has attempted to improve inner laminar shear in such materials through functionalized carbon nanotubes, developed for high performance composites.

The Army and the Department of Energy are interested in magnesium and magnesium alloys. By reducing the grain size and controlling the structure and texture of magnesium at the nanoscale, these materials were made viable for the penetration and the blast resistance that the DoD and Army need. This is a combined experimental and computational effort associated with a very traditional material, which emphasizes the importance of processing, structural modeling, simulation, and experimentation.

In addition to these various projects, the Navy is developing nanoadditives in filtration and biocides in filtration materials, enhancing composite material with CNT and carbon-fiber structures, and investigating multifunctional materials and coatings, highly conductive resins through nanofiber energy materials, photovoltaics, and nanoadditives for body armor.

The Future of Nanotechnology

Some of the technical challenges for materials and processes in the 21st century include:

- Analytical life assessment & prediction (including inspection and prognostic tools)
- Analytical alloy & materials system design
- Accelerated aging & realistic correlation
- Micro/nano-materials properties & characterization, especially physical and mechanical properties
- Practical, applicable optical/photonic materials & processes
- Intelligent, low volume materials processing & manufacture
- Certification of replacement materials
- "Zero" environmental and health impact

For materials research, the following approaches seem promising: self-assembly; tailored surface engineering, both physical & chemical; commodity photonics; nano/micro/meta materials design for function; structure-properties studies at new or newly observable/controllable scales, e.g., quantum domains and cellular micro/macro-structures.

In the future, we will see the continued maturation of nanomaterials, nanomanufacturing, and nanoprocessing. There will be continuing opportunities to realize nanoenabled products. Nanoscale knowledge, understanding research will continue to blend with other disciplines. This diffusion will be great for the science base, specifically the domestic science base and domestic competition.

However, it will have the effect of decreasing the visibility of nanotechnology and science itself. The union of processing research and processing understanding with the materials discovery and materials applications will continue to be important.

References

1. More information is available at the following website: www.defenseinnovationmarketplace.mil/resources/ASD(R&E)_Strategic_Guidance_May_2014.pdf (accessed March 2016).
2. See current and past technical highlights and investment plans in the President's Budgetary Supplement documents: www.defenseinnovationmarketplace.mil/resources/ASD(R&E)_Strategic_Guidance_May_2014.pdf (accessed March 2016).
3. U.S. Naval Research Laboratory. *Institute for Nanoscience*; http://www.nrl.navy.mil/nanoscience/ (accessed March 2016).
4. *Institute for Soldier Nanotechnologies*; http://web.mit.edu/isn/ (accessed March 2016).

Chapter 6

National Institutes of Health: An Introduction to Nanotechnology Funded Research in Biology and Medicine

Lori A. Henderson*

National Institutes of Health, National Cancer Institute,
9609 Medical Center Drive, Rockville, Maryland 20850, United States
***E-mail: hendersonlori@mail.nih.gov.**

The National Institutes of Health (NIH), a part of the U.S. Department of Health and Human Services, is the primary Federal agency for conducting and supporting biomedical research. The NIH's mission is to seek fundamental knowledge about the nature and behavior of living systems and to apply that knowledge to enhance health, lengthen life, and reduce illness and disability. In support of this mission, the NIH has invested about $4.2 billion since 2001 in nanotechnology (NT) research. This investment is leading to fundamental changes in understanding biological processes in health and disease, as well as enabling novel diagnostics and interventions for treating disease. NIH scientists are developing molecular agents and methods for earlier and more accurate diagnosis, therapies aimed directly and selectively at diseased cells, and are exploring root causes of common and rare diseases at the proper length scales. Work is underway to move these research tools and devices into clinical practice and to turn personalized medicine into a reality. This chapter provides a brief introduction to the NIH's NT programs since the inception of the National Nanotechnology Initiative. A more comprehensive description of federally funded R&D and their accomplishments is forthcoming.

Background and History

The NIH conducts and supports research in the causes, diagnosis, prevention, and potential cure of human diseases. The NIH consists of 27 Institutes and Centers. Each of these institutes and centers have their own unique research agenda and strategic plans in support of the NIH's total mission. The agency conducts biomedical research in its own laboratories; supports the research of non-Federal scientists in universities, medical schools, hospitals, and research institutions throughout the country and abroad; helps train the next generation of biomedical research investigators; and fosters communication of medical information. Majority of these institutes are funding extramural NT R&D via specially formulated initiatives or as investigator-initiated grants and contracts. Through these strategic NT initiatives (solicitations), NIH funds basic and clinical research that impact the diagnosis, treatment, or overall management of diseases and patient care with new and emerging ideas.

Moreover, NIH is one of 20 Federal departments and independent agencies that participate in the National Nanotechnology Initiative (NNI) with the goal of understanding and harnessing the unique aspects of nanoscale structures, materials, and assemblies for biomedical research. Consistent with the NNI's definition, NT programs and projects reported by the NIH also include nanoscience and nanoengineering at the molecular and/or macromolecular levels with respect to phenomena and materials at a scale of 1-100 nm and creating devices and systems that have novel properties and functions because of their small size. For instance, nanomedicines exploit the nanoscale manipulation of materials to improve drug delivery. These technologies aim to enable breakthroughs in many disease areas such as cancer.

Though NIH scientists began exploring how to manipulate cells and DNA at the nanoscale as early as 1998, the agency did not dedicate substantial funds to the study of biological processes at the nanoscale until President Clinton launched the NNI in 2001. These resources were aimed at activities that included generating novel and highly effective therapeutics and addressing several challenges in biology and medicine. For example, three bioengineering initiatives were released in 2002 to promote the development and application of NT tools to solve biomedical problems. In 2005, long term opportunities and benefits of NT were fully envisioned when three new initiatives with significant set aside funds were launched. These initiatives stimulated the creation of NT centers and networks supported by the NIH Common Fund (Nanomedicine Roadmap), National Cancer Institute (NCI Alliance Program), and the National Heart, Lung, and Blood Institute (Program of Excellence in Nanotechnology). In addition to encouraging research networking for the development and utilization of tools, each program had specific goals and were organized to link basic research to clinical outcomes in 10 years. These programs later inspired the application of NT in other disease areas via investigator-initiated grants.

Specific NIH Research Programs in Nanotechnology

NIH Common Fund Nanomedicine Initiative

This initiative aimed to transition novel, fundamental discoveries about the precise control and manipulation of subcellular macromolecular assemblies toward preclinical models relevant to specific medical conditions. The two Nanomedicine Initiatives each comprised of a 5 year funding cycle and independent goals. Phase 1 focused on the understanding of how the biological machinery inside living cells is built and operates at the nanoscale. Phase 2 applied this knowledge to re-engineer these structures, develop new technologies that could be applied to treating diseases, and/or leverage the new knowledge to focus work directly on translational studies to treat a disease or repair damaged tissue. To achieve the translation goals, clinical collaborators were added to the teams as part of the Pathway to Medicine concept. The Clinical Consulting Board and program staff work together in assessing clinical decision points and scientific approaches. The program began in 2005 with a national network of 8 Nanomedicine Development Centers. However, in the second half of this 10-year program, the four centers best positioned to effectively apply their findings to translational studies continued receiving support. The Initiative was completed in Fiscal Year 2015. Discover more about the NIH Common Fund activity by visiting commonfund.nih.gov/nanomedicine.

NCI Alliance for Nanotechnology in Cancer

This program pioneered the development and deployment of NT-based diagnostics and therapeutics for cancer. These new technologies should make earlier and more effective disease diagnosis possible and deliver therapies with reduced side effects. The program built a community of researchers dedicated to using NT to advance the fight against cancer. The NCI's extramural Alliance for Nanotechnology in Cancer program (Alliance) was launched in 2004 to promote the application of NT tools and approaches to basic and applied cancer research. Phase 1 & 2 consisted of four components: Centers of Cancer Nanotechnology Excellence (CCNEs), Nanotechnology Platforms for Cancer Research, Nanotechnology Characterization Laboratory, and a Multidisciplinary Research Training and Team Development. In addition, Phase 2 of the Network included awards to promote research independence and training in the cancer nanotechnology field. NCI remains dedicated to maintaining these networks, centers, and educational resources in Fiscal Year 2016 with the start of Phase 3, now a 15 year program. Learn more about the Alliance by visiting nano.cancer.gov.

NHLBI Programs of Excellence in Nanotechnology (PENs)

These programs created multidisciplinary teams to develop and apply NT solutions to the diagnosis and treatment of heart, lung, blood, and sleep disorders. In 2005, Phase 1 began with a focus on developing *in vitro* and *in vivo* diagnostic methods and therapeutic agents for these disorders. Furthermore,

NHLBI awarded contracts to facilitate clinical translation of NT enabled or based products in the second half of this 10 year program. Investigators postulated that the control and manipulation of new materials by precisely configuring atoms and molecules, producing novel molecular assemblies and designing systems of self-assembly to create supramolecular devices on the scale of an individual cell and smaller. The PENs have now concluded. Read more about this program by visiting www.nhlbi.nih.gov.

Comments

In summary, the majority of NIH investment from 2001 – 2015 supported the development of nanodevices and systems with significant contributions in foundational research, and two key program component areas, PCA 3 & 2, respectively, as defined by the NNI to categorize budget investments by the agencies. PCA 3, in particular, focuses on the principles of nanoscale science and engineering to create novel devices or improving existing one. It is important for one to understand that the devices and systems themselves are not restricted to this size but the enabling science and technology must be at the nanoscale. For instance, the Nanomedicine Center for Protein Folding has produced the first single molecule snapshots of the aggregates at the cellular level associated with Huntington disease and subsequently re-engineer chaperonin fragments as therapeutic agents. The Nanomedicine Center for Optical Control was successful in synthesizing new photoswitches that could restore photopic light responses to degenerating mouse retina. See Table 1 for a general desctiption of devices and systems.

Today, NIH-supported researchers are employing novel engineering strategies that will enable breakthroughs in many areas (see Table 2), including precisely engineered materials to develop novel therapies and devices. These materials aim to reduce drug toxicity and dose, enhance targeted delivery, attack disease before clinical symptoms develop, repair or replace failing organs, and control the spread of disease or infections with cost-effective nano-enabled technologies. Examples include the first rapid diagnostic test to detect Dengue, Yellow Fever, and Ebola and the delivery of "nanoaerosols" to the airways of infants who are ill. As of 2015, more than 100 investigational new drug applications (INDs) were submitted to the FDA for drug products containing nanomaterials (in consultation with FDA). Of those, approximately 50 are listed as "active" although this is not necessarily an indication of ongoing clinical trials (see Table 3). Because of the abundance of studies, upcoming publications and NNI Accomplishment Reports will be used to highlight examples of the impact on understanding biology of diseases and specific clinical applications pertaining to medical diagnosis, treatment, and disease management.

Table 1. Examples of NT-Enabled Devices, Systems, and Platforms

NIH portfolio encompasses the development of:

- ❑ Nanodevices to identify and diagnose disease by finding biomarkers and pathogens to detect disease early
- ❑ Multifunctional nanoparticles capable of targeting the right amount of therapeutics to the exact location in the body where they are needed & monitoring treatment effectiveness
- ❑ Nanoparticle tools that extend beyond the limits of current technologies to reveal the biomolecular basis of diseases by transducing molecular binding events into optical images & providing real time observation of distinct cellular processes at the nanoscale
- ❑ Antimicrobial coatings and nanomaterials to control the biotic-abiotic interface for restoring vision and hearing, better implants to repair tissue damage
- ❑ Nanofluidic platforms to produce novel imaging probes to detect diseases, monitor immune system function, screen nanoparticles for toxicity
- ❑ Engineered scaffolds to regenerate bladder and other organs, repair damage tissue

Table 2. Sampling of Technology Focus Areas Involving Nanoengineering

GOAL

- ❑ Detecting Disease *Before* Health has deteriorated (symptoms)
- ❑ Repair or replace damaged body parts
- ❑ Improve diagnostic and therapeutic approaches

APPLICATIONS

<u>Drug and Gene Delivery Systems</u>

Nanoparticles

- Controlled Release
- Targeted Delivery
- Imaging & Monitoring Activity

<u>Medical Devices and Sensors</u>

Nano Devices, Arrays, and Switches

- Biochemical detection
- Sensing biomolecules
- Cell-based Assays

<u>Imaging Modalities</u>

Contrast Agents & Probes

- Image cell signaling processes
- Cell tracking & monitoring intracellular events
- Microscopic detection
- Nanoscale chemical Imaging
- Intra-operative surgery

<u>Tissue Engineering</u>

Nanostructures

- Physical cues for cell signaling & mechanics
- Nanopatterned surfaces to control cell response & fate
- Nanoengineered surfaces to control topography & presentation of ligands

Table 3. Selected Number of Areas of Active Clinical Research and Development

GOAL
NANOTECHNOLOGY FOR MEDICAL DIAGNOSTICS AND TREATMENT
• Early diagnosis using *in vitro* assays and devices or *in vivo* imaging techniques • Multifunctional nano-therapeutics and post-therapy monitoring tools • Devices and techniques for cancer prevention and control • Devices and techniques to characterize physiochemical properties & biological interactions for environmental health and safety (safe delivery)
CLINICAL APPLICATIONS
ONCOLOGIC, CARDIOVASCULAR, INFECTIOUS, & NEUROLOGICAL DISEASES

Closing Remarks

In conclusion, the full potential of NT in medicine remains to be seen as recent discoveries are in pre-clinical or early clinical testing. NT also has the potential to meet goals, address challenges, and provide solutions to current Administration Implemented activities, such as the Precision Medicine Initiative. Launched in 2015, the federal-wide Precision Medicine Initiative aims to generate the scientific evidence needed to move the concept of precision medicine into every day clinical practice. Precision medicine is an emerging approach for disease treatment and prevention that accounts for individual variability in genes, environment, and lifestyle for each person. Current initiatives are underway at the National Cancer Institute, for instance, to significantly expand efforts in cancer genomics to create prevention and treatment successes for more cancers. While significant advances in precision medicine have been made for select cancers, the practice is not currently in use for most diseases.

The NIH will continue to work with our NNI partner agencies, as part of its mission, to build "*a future in which the ability to understand and control matter at the nanoscale leads to a revolution in technology and industry that benefits society.*" Please contact Dr. Henderson if you are interested in knowing more about NIH activities in nanotechnology. She currently serves as the Agency co-Chair of the NSET Subcommittee, a Subcommittee of the National Science and Technology Council's Committee on Technology, under the White House Office of Science and Technology Policy. She is also a Program Director in the Clinical Trials Branch of the Division of Cancer Treatment and Diagnosis at the NCI.

Acknowledgments

I would like to thank the NIH funded scientists and clinicians, in addition to the NIH Program Staff for the NDC, Alliance, and PENs, for their remarkable insight and dedication to the field. For their inventiveness, dedication, and commitment has contributed significantly to the NNI's goal on "Advancing a World-Class Nanotechnology Research and Development Program for the nation. In addition, I would like to thank Dr. Francis Collins, Director of the NIH, and his staff within the Office of Science Policy, for their contributions to this summary and support for my role as the Agency co-Chair of the NSET Subcommittee.

Chapter 7

University Roles in Economic Development

Michael Liehr*

College of Nanoscale Science and Engineering, SUNY Polytechnic Institute, 257 Fuller Road, Albany, New York 12203, United States
***E-mail: MLiehr@sunycnse.com.**

The classic approach of universities to economic development is via creation of spin-off companies based on the intellectual property developed by its faculty and students. The obstacles to overcome in this endeavor are primarily related to fostering a commercialization-based environment and offering funding opportunities to bridge the gap between research and a commercial product. The latter often involves engaging venture capital firms or angel investors.

A second approach implies a close collaboration between universities and state governments, enabling the settlement of established companies at or close to campuses, typically via science or industrial parks. Here, physical proximity of commercial enterprises allows the use of university infrastructures as well as the training and recruitment of specialized skills needed by the companies.

At SUNY Polytechnic Institute (SUNY Poly), the second approach has been used extensively with an initial focus on nano-electronics and will be described using specific consortia examples. The first approach, for SUNY campuses, is facilitated by the Research Foundation for SUNY as described below.

SUNY Startups

A university serves to create and bring about new ideas. Sometimes companies are formed around these new ideas generated by universities.

The SUNY Polytechnic Institute is part of the SUNY system, the largest comprehensive state university system in the country, with 64 campuses, 88,000 faculty and staff, 468,000 students and about 3 million alumni. Its size is due, in part, to the inclusion of community colleges, technology colleges, comprehensive colleges, and doctoral degree granting institutions all part of one system. An organization called the Research Foundation for SUNY serves as the administrative of SUNY campuses for externally sponsored research and technology transfer. Total sponsored program expenditures across all campuses through the Research Foundation is close to $1 billion per year.

From a technology transfer perspective, preliminary data shows that since 2011 the system has had 1,846 faculty, staff, and students be named as inventors on 1.410 new technology disclosures, 301 disclosures were included in 244 commercialization agreements with industry partners, and 1,183 new U.S. patent applications were filed.

To date, SUNY has 77 active startups that were formed to commercialize SUNY-developed innovations, over 160 companies in 17 business incubators, and 89 companies with Small Business Innovation Research/Small Business Technology Transfers (SBIR/STTR) awards subcontracted to SUNY, as well as 6 state-wide Networks of Excellence. These numbers are not that impressive, given the number of people involved in the SUNY system, and in comparison with universities in California and Massachusetts. A large part of this is due to funding leaving the state rather than remaining in the state. There is an intense focus in the SUNY system to improve those numbers.

The governor of New York has taken a very active role in fostering the collaboration between venture capitalists and higher education, and examining barriers that slow or thwart commercialization efforts. He helped created START-UP NY, which allows anyone starting a business to work in a tax-free environment for 10 years, waiving all taxes except federal taxes. He also focuses a lot on innovation, but it will take a lot longer for that to take hold – the innovation support is less than a year in development.

One of the things SUNY is trying to do is to emulate a model that has been very successful on the west coast. SUNY has been working with Semi, a trade association focusing on equipment supply in the electronics industry, in the semiconductor and packaging space. It is composed of about 2,000 member companies. It brings together venture capitalists and startups. Right now, there is a lot of venture capital in New York City, but not many companies taking advantage of the funding available. Eventually, SUNY would like to expand these operations beyond New York City.

SUNY Public-Private Partnerships

The National Science Foundation (NSF) has specifically mentioned SUNY Polytechnic Institute (SUNY POLY) as an organization that is successful in public-private partnerships. SUNY POLY has a state-wide presence. This reflects the position the SUNY president holds within the state, as the senior economic advisor to the New York governor. While New York City is a huge economic center, upstate New York is not doing as well, economically speaking. A lot of industries, such as textiles, have left the state, so the SUNY POLY state-wide sites are an effort to boost economic growth in these upstate areas (Figure 1), with a focus on high tech. It is aimed at creating a high-tech entrepreneurial ecosystem in upstate New York. The governor has invested a substantial amount of money into this.

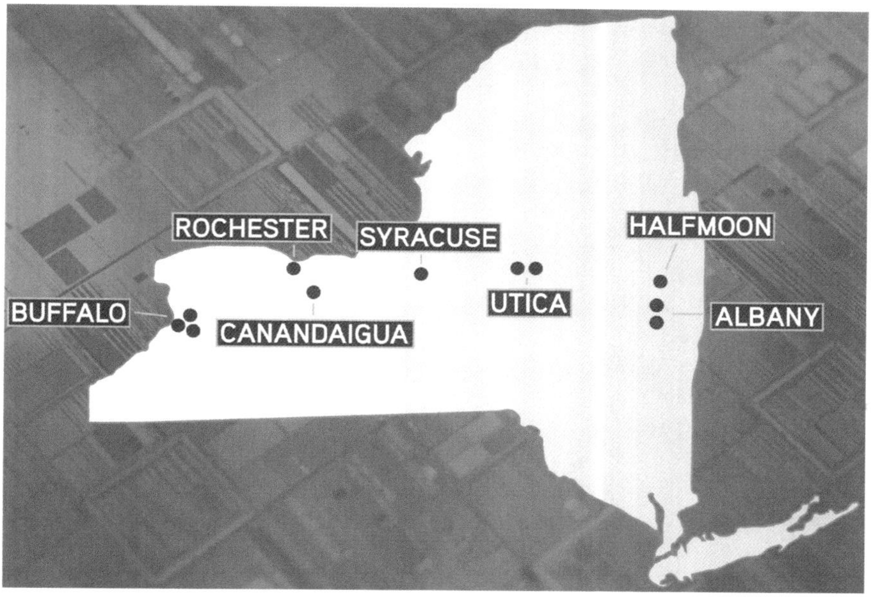

Figure 1. Locations of SUNY POLY sites across the state of New York.

SUNY Facilities and Programs

SUNY Poly's Albany NanoTech Complex, home to the Colleges of Nanoscale Science and Engineering, is a fully-integrated research, development, prototyping, and educational facility (*1*). The Albany site is primarily focused on silicon semiconductors. There is 137,000 square feet of clean room space, which is the size of a modern semiconductor factory, making it the largest facilities in the country. The closest facility to this in size is IMEC in Belgium. A lot of the buildings at the Albany site are 300-mm silicon wafer manufacturing buildings, which focuses on research and development (*2*). There are also 450-mm tools on the site, accommodate the needs of larger companies such as Intel, Samsung, and AT&T.

The 300-mm wafer facilities are equipped with complete toolsets, allowing SUNY POLY to manufacture semiconductors at the most advanced level. The Albany site is also the only facility that has 450-mm tools. The facility features leading edge lithography to support lithography development and help manufacture the semiconductors. The SUNY POLY line is backwards compatible to 65 nm – used to ease the transition to different sized wafers in semiconductor production, and ensure a similar yield between different sizes. This line is capable today of 7-nm litho generation (*3*). The facility has a capacity of roughly 5,000 wafers being injected into the line, known as "short loops" – which is less than what a factory would use, due to the research-focused nature of the production. There are also a fairly large number of on-site equipment companies.

The Albany location has SEMATECH on-site, now headquartered in Albany and serving as a global research, development, and economic vitality arm of the State of New York, through SUNY Poly. Their claim to fame is bringing back the semiconductor industry to the U.S. The location brings together competing interests, such as IBM, Intel, and Samsung, through stratification. Stratification makes these competing interests focus on the "recipes" used in the semiconductor manufacturing, rather than building the manufacturing equipment themselves. Equipment companies manufacture the tools, which everyone can then buy. It has made a big difference, because companies can now focus on their market niche rather than production. This represents a concept of pre-competitive collaborative work, where the focus is on the semiconductor manufacturing and equipment. People get together and exchange best practices in manufacturing, and get feedback from others who are willing to contribute. It has been a successful model, given the number of companies that SUNY POLY has been able to attract.

SUNY POLY is expanding its model to other fields and industries, where people can use their knowledge, equipment and people to work in adjacent spaces with others to collaborate on ideas.

GE and SUNY POLY have teamed up as a consortium to apply the practices in the silicon semiconductor space to the silicon carbide space to create additional options for products and revenue.

At the Utica, NY site, they are taking advantage of 3D packaging. They have built a facility with about 95,000 square feet of clean room that will be the home of a packaging consortium, where the focus is on packaging production. The automation involved in this production allows the U.S. and Europe to compete with China, despite the lower wages China pays its manufacturing employees – especially given that this is low-cost packaging. The advanced packages developed at Utica constitute a concept SUNY believes they can bring to the U.S. and keep in the U.S.

Pending right now is a proposal for an integrated photonics program through DOD's Funding Opportunity Advancement. It is the largest proposal brought forth for FOA. The proposal asks for $110 million spread over five years. This is a technology where you are bringing light onto a chip. Light on chips addresses a lot of problems associated with the standard chips used today, which use wires, switches, and transistors. There are a number of industry segments interested in participating in this. The proposal needs to fit within the interests of the Department of Defense, as per the FOA guidelines, but the healthcare industry generally has also taken an interest in using this technology in the medical field.

References

1. *SUNY Poly Colleges of Nanoscale Science and Engineering (CNSE)* (website); http://www.sunycnse.com/WorldClassResources/StatewideFacilities.aspx (accessed October 30, 2015).
2. *SUNY Poly CNSE Wafer Processing Research & Development* (website); http://www.sunycnse.com/WorldClassResources/CNSEWaferProcessingRandD.aspx (accessed October 30, 2015).
3. Anonymous. *Researchers produce industry's first 7-nm node test chips*; http://phys.org/news/2015-07-industry-7nm-node-chips.html (accessed October 30, 2015).

Health, Safety, Environmental, and Regulatory Issues

Chapter 8

Regulations: Facilitating Advancement or Serving as a Barrier - A Shared Responsibility

Terry L. Medley*

DuPont Product Stewardship and Regulatory, Chestnut Run Plaza 708-110, 974 Centre Road, P.O. Box 2915, Wilmington, Delaware 19805, United States
***E-mail: terry.l.medley@dupont.com.**

Product innovation and sustained economic growth for science based companies requires the continued integration of diverse technology capabilities to create high performance, improved products that meet market needs and demands. Nanotechnology is an enabling technology with applications applicable to every materials market (e.g., medical applications, devices and microelectronics, energy/industrial, bioremediation, coatings and dispersions). The responsible utilization and adoption of this technology will be impacted either favorably or unfavorably by the regulatory requirements for its oversight and further development. These regulatory requirements can either provide a platform which facilitates decision making and responsible technology adoption or create unnecessary barriers to innovation and technology utilization. Regulations can assist in identifying potential risks while avoiding unnecessary data generation, time delays and increased costs. Ensuring that regulatory oversight addresses potential environmental, health and safety (EHS) concerns while allowing innovation and technology adoption is best achieved through the use of an integrated approach. Such an approach should utilize, as appropriate, product stewardship, voluntary, collaborative and mandatory measures to achieve the desired EHS objectives. To have the best chance for success, such an integrated approach must be a shared responsibility among all stakeholders. This paper will explore three key aspects of such a shared integrated approach to efficient, effective and proportional regulatory oversight of nanotechnology.

Introduction

Nanotechnology comprises the creation of functional materials, devices, and systems through control of matter on the nanometer (1 to 100 nm) length scale and the application of novel properties and phenomena developed at that scale (*1*). At the nanoscale, materials can be created with valuable chemical and physical properties which can enable a wide range of technological applications (*2*).

The National Nanotechnology Initiative (NNI), established in 2001, is a U.S. government research and development initiative involving 20 Federal departments, independent agencies and commissions working together toward the shared and challenging vision of a future in which the ability to understand and control matter at the nanoscale leads to a revolution in technology and industry that benefits society. Since the inception of the NNI, participating agencies have invested more than $22 billion in funding (including the President's 2016 budget request) fundamental applied nanotechnology R&D; world-class characterization, testing, and fabrication facilities; education and workforce development; and efforts directed at understanding and controlling the environmental, health, and safety (EHS) aspects of nanotechnology (*3*).

The rationale for the U.S. government's significant investment in furthering the development of nanotechnology is clear. In a letter to Congress supporting the 2016 budget request, Office of Science and Technology Policy Director, John Holdren stated, "The proposed NNI budget for Fiscal Year 2016 of $1.5 billion will continue to advance our understanding of nanoscale phenomena and our ability to engineer nanoscale devices and systems that address national priorities and global challenges… Continuing U.S. leadership in innovation will be enabled by nanotechnology and other emerging technologies. Sustained support for the NNI is vital to providing the transformational knowledge and technologies that can benefit society and create the businesses, jobs, and opportunities of the future" (*4*).

The proposed fiscal year 2016 NNI budget also allocates 7.1 percent of the $1.5 billion budget to address environmental, health and safety aspects. The government's nanotechnology spending for EH&S has steadily increased from 2.8 percent in fiscal 2006 budget . This increase reflects a commitment to address both potential benefits and any potential safety concerns (*5*). Although support funding is essential, responsible technology utilization and adoption will also be impacted by the regulatory requirements for its oversight and further development. Such regulatory requirements can either provide a platform which facilitates sound decision making and responsible technology adoption or create unnecessary barriers to innovation and technology utilization.

Ensuring that regulatory oversight addresses potential environmental, health and safety (EHS) concerns while allowing innovation and technology adoption is best achieved through the use of a shared intergrated approach. Such an approach will address environmental, health and safety concerns and facilitate economic growth and innovation. It seeks balance - to neither over or under regulated adoption of the technology.

This shared intergrated balance approach to the responsible development of nanotechnology has three key components; industry product stewardship commitments; effective collaborations to address scientific uncertainty and EHS concerns; and risk based regulatory principles, policies and regulations.

Industry Product Stewardship Commitment

Product Stewardship comprises understanding, controlling, and communicating a product's environmetal, health , and safety related effects throughout its life cycle, from production (or extraction) to final disposal or reuse (*6*). Within my company we further define Product Stewardship as the shared responsibility for identifying, managing, and communicating product health, safety and environmental information and issues along the entire value chain; and adding value and meeting customer, market, societal, and stakeholder expectations throughout the product life cycle. And it is in support of our company's right to operate, innovate and compete (*7*).

We are fully aware that utilization of new technologies also raises questions about unknown or unintended environmental, health and safety concerns. We are committed to good product stewardship practices for all products that we develop regardless of the technology used for the product's development.

On June 24, 2005 the Wall Street Journal ran an article titled "Let's Get Nanotech Right". This article was co-authored by then DuPont CEO Chad Holliday and the President of Environmental Defense (now Environmental Defense Fund), Fred Krupp. It highlighted that, "Good product stewardship requires a commitment to identifying and managing any potential risks. And that an early and open examination of the potential risks of a new product or technology is not just good common sense—it is a good business strategy. With the right mix of voluntary corporate leadership, coordinated research, and informed regulation, we can reap the benfits of this promising technology while reducing the likelihood of unintended consequences" (*8*).

In September 2005, DuPont and Environmental Defense Fund agreed to collaborate on a framework for the responsible development, production, use and disposal of nanoscale materials. The purpose of this framework was to define a systematic and disciplined process that can be used to identify, manage and reduce any potential health, safety and environmental risks of nanoscale materials across all lifecycle stages. The Nano Risk framework that followed was released to the public on June 21,2007 and also adopted internally by DuPont (*9*).

The DuPont-Environmental Defense Fund Nano Risk Framework

Environmental Defense Fund, an environmental advocacy organization, and DuPont, a science company developed a comprehensive, practical, and flexible framework for evaluating and addressing the potential risks of nanoscale materials. The framework defines a systematic and disciplined process for identifying, managing, and reducing any environmental, health, and safety risks of engineered nanomaterials across all stages of a product's lifecycle. The framework offers guidance on the key questions an organization should consider in developing applications of such materials, and on the key information needed to make sound risk-evaluation and risk-management decisions. The framework allows users to move ahead despite areas of incomplete or uncertain information, by using reasonable assumptions and by compensating for knowledge gaps with appropriate risk-management practices. Furthermore, the framework describes a system to guide information generation and update assumptions, decisions, and practices with new information as it becomes available. The framework also offers guidance on how to communicate information and decisions to key stakeholders.

A Mix of Familiar and New Elements - Users acquainted with other risk-management frameworks will recognize some familiar elements here, but it also incorporates new elements. Central to the framework is the development of informational profiles ("base sets") — relevant to the properties, hazards, and exposures associated with a given nanomaterial and its application - for evaluating risks and guiding decisions. The framework is thus information-driven. The framework does not implicitly assume the risk or safety of any material. Where there is little or no information to guide decisions on the potential for a particular hazard or exposure, the framework suggests using "reasonable worst-case assumptions" -or, alternatively, using comparisons to other materials or processes that have been better characterized - along with management practices appropriate to those options. The framework encourages replacing assumptions with real information, especially as a product nears commercial launch, and refining management practices accordingly.

Comprehensive, Flexible and Practical - Since the beginning of the partnership, a key goal of both organizations was to develop a systematic and disciplined process that is comprehensive, flexible and practical. We developed the Nano Risk Framework to be comprehensive enough to thoroughly consider a wide range of potential environmental health and safety issues, throughout the lifecycle of a nanomaterial and its application in a product. We built in flexibility so it can apply to a wide range of nanomaterials and applications, at various stages of development, and in a wide variety of settings, from large companies to small startups. And we designed it to be practical so that it complements many of the product development and product stewardship practices that many companies already have in place.

Nano Risk Framework – This framework has a six-step process, as shown in Figure 1:

Step 1: Describe Material & Application: a basic description, as a way to set up the more thorough look into the material's properties, hazards and exposure throughout its product lifecycle. Users should be able to complete this with information already available in-house or in the scientific literature. The product's lifecycle can be thought of as encompassing all the processes and activities that occur from initial extraction of the material (or its precursors) from the earth to the point at which any of the material's residuals are returned to the environment.

Step 2: Profile Lifecycle(s): The user takes into account the nanomaterial's full lifecycle to develop profiles of the material's key properties, hazards, and exposures, and how those characteristics may change during the material's lifecycle. The properties profile characterizes a nanomaterial's physical and chemical properties. The hazard profile characterizes the nanomaterial's potential safety, health, and environmental hazards. And the exposure profile characterizes the opportunities for human or environmental exposure to the nanomaterial — including exposure both through intended use and by accidental release.

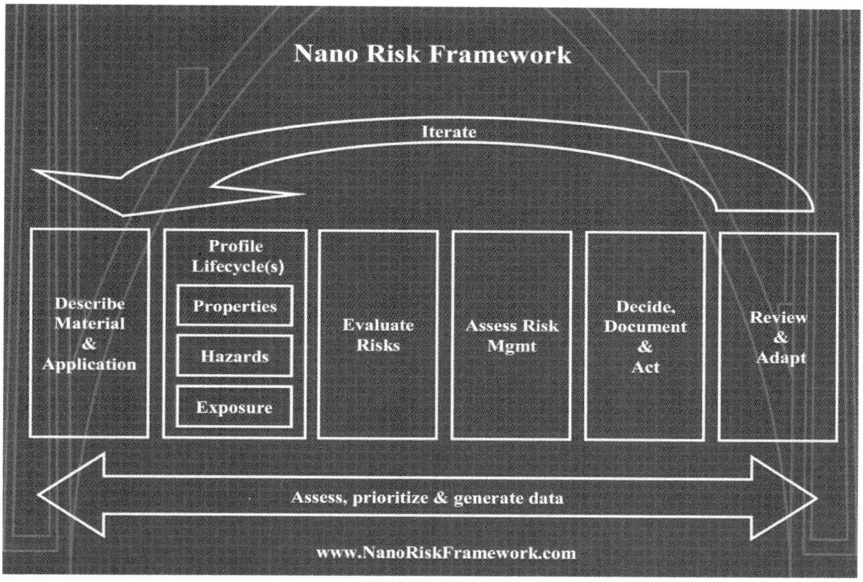

Figure 1. Nano Risk Framework.

Step 3: Evaluate Risks: In this step, all the information generated in the profiles is reviewed in order to identify and characterize the nature, magnitude, and probability of risks presented by a particular nanomaterial and its anticipated application. Here the user considers gaps in the lifecycle profiles, and determines how to address them — either by generating data or by using "reasonable worst case" assumptions or values in place of data.

Step 4: Assess Risk Management: Here the user evaluates the available options for managing the risks identified in Step 3 and recommends a course of action. Options include engineering controls, protective equipment, risk communication, and product or process modifications, or in some cases, halting the development of the material or application.

Step 5: Decide, Document & Initiate Action: In this step, a cross-functional review team reviews the risk evaluation and risk management recommendations and decides whether or in what capacity to continue development and production. The user documents those decisions and their rationale and shares appropriate information with the relevant stakeholders. The user may also decide that further information is needed and initiate action to gather it.

Step 6: Review and Adapt Ongoing Management: This is an iterative process of regularly scheduled reviews as well as event-triggered reviews to update the lifecycle profiles and risk evaluation. This ensures that risk management systems are working as expected, and adapts those systems in the face of new information or new conditions.

EACH STEP requires you to assess what you know and don't know about the material and the application, prioritize your data needs, and take action to address those needs. Throughout the product lifecycle, the process should be revisited and updated, and we expect that as a product moves through the development process, you would require a more thoroughly developed set of information – early R&D wouldn't require as much data as you'd want before launching a large-scale commercial application.

Implementing this Framework will provide users with a comprehensive, effective, and flexible system for addressing the potential environmental, health, and safety risks of nanomaterials and their applications. The Framework offers flexibility in addressing data-generation needs. Hazard information may already be available in the literature for some nanoparticles or applications, or it may be possible to develop such data by "bridging" to other, better-characterized materials. The amount of information required in the Framework is directly related to the potential extent and degree of exposure of the specified application.

The Framework is most effective when incorporated into or paired with a system to ensure its execution. That system may be an existing product-development or product-stewardship process, or it may be a new system designed specifically to implement the Framework. Since establishment of the Framework website the site has been visited over 29,097 times and the document downloaded 12,804 times by individuals representing 146 countries (*10*).

In sum, adoption of the Framework demonstrates commitment to product stewardship and promotes responsible development of nanotechnology products . It also seeks to contribute to the development of science-based policy for nanotechnology safety. For example the International Organization for Standardization(ISO) utilized much of the structure and content from the Framework to develop its 2011 Nanomaterials Risk Evaluation document (*11*).

Efffective Collaboration To Address Scientific Uncertainty and EHS Concerns - OECD WPMN

New technologies by their very nature of being new will raise questions concerning incomplete data, unintended or unkown health and environmental consequences. So it is not a surprise that new nanoscale materials are raising questions about their potential impact on health, safety and the environment. These questions are being widely discussed and considered by regulatory agencies, public and private special interest organizations and in numerous industry, scientific, national and international forums. Such broad involvement and discussion is extremely important.

All stakeholders with an interest or stake in the responsible development and use of new nanoscale materials should work together to allow this technology adoption to reach its full potential. We should seek and advocate for effective collaboration in the development of responsive safety standards and test methods;

coordination of research to generate reliable, peer reviewed data based on sound science and the adoption of appropriate and proportional regulations as needed.

One organization that has a long history of providing a forum for effective collaborations on emerging scientific and other policy matters is the Organisation for Economic Co-Operation and Development (OECD). The OECD was established in 1961 and part of its mission is to provide a setting where governments compare policy experiences, seek answers to common problems, identify good practices and coordinate domestic and international policies (*12*).

For more than 40 years OECD has had a key role in the safe use of chemicals and the protection of human health and the environment. As a part of its response to emerging issues, OECD identified the need to analyze the potential safety concerns caused by manufactured nanomaterials. OECD concluded that although there was much knowledge regarding possible health and environmental effects of traditional chemicals, it was not clear if this knowledge could be transferred directly to nanomaterials. Therefore, OECD held a special session in 2005 on the Potential Implications of Manufactured Nanomaterials for Human Health and Environmental Safety. At this meeting it was agreed to set up a program of work to assist countries in the implementation of national policies for the responsible development of nanotechnologies (*13*).

The Working Party On Manufactured Nanomaterials (WPMN) was estabished at OECD as a Programme on the Safety of Manufactured Nanomaterials in 2006. The objective of the WPMN is to promote international co-operation in human health and environmental safety related aspects of manufactured nanomaterials in order to assist in the development of rigorous safety evaluation of nanomaterials.

The WPMN brings together representatives from the OECD member countries, observer countries, industry, trade unions, Environmental NGOs and other international groups to share their knowledge , experiences and coordinate research to address scientific uncertainty , data gaps, evaluation of existing safety standards and test methods and the generation of data to support regulatory decision making . It is the intent of the program to address all of the different components needed for thorough risk assessments for human health and the environment (*14*). Its program of work over the last decade has included development of the OECD database on EHS research; of guidance on exposure measurement, exposure mitigation, and alternative test methods; co-operation on risk assessments; a series of workshops and expert meeting on key EHS issues; and a Sponsorship Programme for the testing of manufactured nanomaterials (*15*).

The sponsorship testing program was an exploratory testing program aimed at testing specific nanomaterials for their physical chemicals properties, environmental fate and behavior, ecotoxicity and toxicity using appropriate testing methods. Representative manufactured materials were selected that were either in commerce or expected to be in commerce in the near future. The OECD is now going through the process of making the results of the sponsorship testing program available. Other major contributions from the WPMN have included development of guidance documents; updated and new test guidelines; findings from expert workshops; exposure assessments and the project on nanomaterials in tires (*16*).

The OECD and its member countries concluded that the approaches for testing and assessment of traditional chemicals are in general appropriate for assessing the safety of nanomaterial , but may have to be adopted to the specificities of nanomaterials (*17*). There appears to be a growing global consensus that existing laws and regulatory constraints are sufficiently robust to regulate the safety of nanoscale materials (*18*). On September 19, 2013 the OECD recommended that members, to manage the risks of manufactured nanomaterials, apply the existing international and national chemical regulatory frameworks or other management systems, adapted to take into account the specific properties of manufactured nanomaterials (*19*).

Risk-Based Regulatory Principles, Policies, and Regulations

Managing the oversight of nanomaterials with existing international and national chemical regulatory frameworks will be a challenge. It will be a challenge to our capacity for balancing technological progress with protection of health and the environment. Balanceing a rapidly advancing technology's ability to offer potential for social good with uncertainty regarding health and environmental concerns (*20*). Regulating nanomaterials simply because of its size or its manufacturing process will discriminate improperly against nanomaterials unless as a class they are more risky than non-nanomaterials. And this has not been established (*21*). One way to help avoid this undesireable outcome is to adopt risk based regulatory principles, policies and regulations.

On January 18,2012 President Obama signed Executive Order 13563, Improving Regulations and Regulatory Review. It stated that "Our regulatory system must protect public health, welfare, safety, and our environment while promoting economic growth, innovation, competitiveness, and job creation. It must be based on the best available science. It must allow for public participation and open exchange of ideas. It must promote predictability and reduce uncertainty." The executive order asked that agencies use the best, most innovative and least burdensome tools for achieving regulatory ends. It also sought to ensure that regulations are accessible, written in plain language and easy to understand (*22*).

On June 9, 2011, as a follow-up to the executive order, a memorandum to heads of executive departments and agencies was issued entitled " Policy Principles for the U.S.-Decisions Making Concerning Regulation and Oversight of Application of Nanotechnology and Nanomaterials". The memorandum concluded that existing regulatory statues provide a firm foundation for the oversight of nanomaterials and outlined several key principles essential to a risk based approach. It instructed the executive departments and agencies to seek to use the best available science evidence and to take into account any unique properties and behaviors associated with nanomaterials. It also cautioned that , " nanomaterials should not be deemed or identified as intrinsically benign or harmful in the absence of supporting scientific evidence, and regulatory action should be based on such scientific evidence" (*23*).

Conclusion

Enabling technology innovation and application, while identifying and addressing environmental health and safety (EHS) concerns requires an oversight system that utilizes all viable options in an integrated manner. Such an intergrated approach should utilize, as appropriate workplace safety and product stewardship initiatives, voluntary and mandatory agency actions and collaborative risk frameworks and data generation efforts. These industry initiatives, agency actions and collaborative efforts should be based on science/risk-based policies. This paper has highlighted one company's commitment to product stewardship; development of a nano risk framework for responsible development ; a successful OECD collaborative effort to address scientific uncertainty and U.S Government's issuance of science risk based regulatory policies and principles. There are similar activities occurring in many other jurisdictions . However, to truly enhance our chances for success, such an integrated approach must also be viewed by industry, interested stakeholders and regulatory agencies as a **shared responsibility and a societal obligation**.

References

1. *ACS Presidential Symposium on Nanotechnology*; Denver, CO, March 2015.
2. *Implications of Nanotechnology for Environmental Health Research*; Goldman, L., Coussens, C., Eds.; Institute of Medicine (U.S.) Roundtable on Environmental Health Sciences, Research, and Medicine; National Academies Press: Washington, DC, 2005; Chapter 2 (The Promise of Nanotechnology).
3. National Science and Technology Council, Committee on Technology, Subcommittee on Nanoscale Science, Engineering, and Technology. *The National Nanotechnology Initiative, Supplement to the President's 2016 Budget,* Washington, DC, 2015, p. 7.
4. Holdren, J. Letter to Congress in ref. 3.
5. National Science and Technology Council, Committee on Technology, Subcommittee on Nanoscale Science, Engineering, and Technology. *The National Nanotechnology Initiative, Supplement to the President's 2016 Budget*; National Science and Technology Council: Washington, DC, 2015; p 23.
6. Business Dictionary.com. *Product Stewardship definition*; www.businessdictionary.com (accessed on 12/9/15).
7. Personal conversation with James Romine, DuPont VP Product Stewardship and Regulatory.
8. Holliday, C.; Krupp, F. *Lets get Nanotech Right. Wall Street J.* **2005** June14.
9. Environmental Defense Fund – DuPont Nano Partnership. *Woodrow Wilson Institute Public Meeting - Release of the Nano Risk Framework*; June 21, 2007.
10. Environmental Defense Fund – DuPont Nano Partnership. *Nano Risk Framework*, 2007; http://qsinano.com/wp-content/uploads/2014/05/nano_risk_framework_dupont.pdf (accessed on 1/12/16).

11. Maynard, A. *International Standards Organization guidelines for evaluating nanomaterial risks – are they any good?* 2020 Science, May 26, 2011; www.2020science.org (accessed on 12/9/15).
12. *Organisation for Economic Co-Operation and Development*; www.oecd.org (accessed on 12/9/15).
13. Organisation for Economic Co-Operation and Development. *Six Years of OECD Work on the Safety of Manufactured Nanomaterials*; September 2012; www.oecd.org (accessed on 12/9/15).
14. Organisation for Economic Co-Operation and Development. *Six Years of OECD Work on the Safety of Manufactured Nanomaterials*; September 2012; p 3.
15. Clancy, S. F. *NanotechnologyActivities at ISO and OECD*; Presentation delivered at Global Chem Meeting, March 5, 2005; slides 9, 11, 12.
16. Clancy, S. F. *NanotechnologyActivities at ISO and OECD*; Presentation delivered at Global Chem Meeting, March 5, 2005; slides 7, 8.
17. Organisation for Economic Co-Operation and Development. *Six Years of OECD Work on the Safety of Manufactured Nanomaterials*; September 2012; www.oecd.org (accessed on 12/9/15).
18. Bergeson, L. L. *Emerging Global Nanotechnology Regulatory Developments*; Presentation delivered at DuPont Nano-Summit, February 10, 2015; slide 7.
19. Organisation for Economic Co-Operation and Development, *Recommendation of the Council on the Safety Testing and Assessment of Manufactured Nanomaterials*; September 19, 2013; www.oecd.org (accessed on 12/9/15).
20. Malloy, T. F. *Soft Law and Nanotechnology; A Functional Perspective*, http://ssrn.com/abstract=2002347 (accessed on 12/9/15).
21. Marchant, G. E.; Sylvester, D. J.; Abbott, K. W. A New Soft Law Approach to Nanotechnology Oversight: A Voluntary Product Certification Scheme. *UCLA J. Environ. Law Policy* **2010**, *28*, 131.
22. *U. S. Executive Order 13563*, January 18, 2011.
23. *Policy Principles for the U.S. Decision Making Concerning Regulation and Oversight of Applications of Nanotechnology and Nanomaterials*; Memorandum for the Heads of Executive Departments and Agencies, June 9, 2011.

Chapter 9

Sustainability and Life-Cycle Issues

Jeffrey Morris*

Office of Pollution Prevention and Toxics, Environmental Protection Agency, Washington, DC 20460, United States
***E-mail: morris.jeff@epa.gov.**

U.S. EPA's Office of Pollution Prevention and Toxics has over 10 years of experience engaging with the regulated community, scientists, the international community, and others on the regulation of manufactured nanomaterials. As of November 2014, EPA had made regulatory decisions on over 160 nanomaterials for which EPA had received premanufacture notification. This article will discuss the lessons EPA has learned in evaluating nanomaterials with respect to environmental sustainability across the life cycle, and what challenges remain in developing and applying scientific information to determine the safety of manufactured nanomaterials.

Regulatory Decision Making

Numerous decisions go into developing nanomaterials. Clearly, most of those decisions center on material performance and the business case for incorporating a nanomaterial into a product. In addition to those considerations, to advance the development of environmentally sustainable nanomaterials, a life cycle perspective is needed. For regulators, the decisions that go into the design of products and molecules have implications for chemical safety. When the regulatory science community looks at metrics of environmental sustainability for nanomaterials, it is useful to understand the decisions made by designers, fabricators, and manufacturers when they consider what properties to incorporate into their materials.

In general, applying life-cycle assessment to nanomaterials will require considerably more information than we currently have because, to date, the scientific community has incomplete information on what attributes of nanomaterials drive their behavior in air, water, land and biological systems. When the regulatory science community looks at a particular material type, it typically faces uncertainty about how the nanomaterial's design features will affect its movement and behavior in the environment.

If nanomaterial designers do not, or are not able to, apply life-cycle thinking to how their nanomaterials will be produced, used, and ultimately disposed of or recycled, then opportunities will be missed to incorporate green chemistry and environmental sustainability principles into the development and application of nanotechnology.

Likewise with applying life cycle thinking in regulatory decision making, regulators are concerned not only about toxicity or exposure potential at the point of use or introduction into the environment. They also consider how the nanomaterial is going to be produced and handled by humans when it is used, disposed of, and recycled, as well as how it will change as it moves through air, water, or soil. All of these parts of the life cycle are important when considering whether or not something is environmentally sustainable.

Life-cycle assessment approaches have to be useful and usable for both the regulated community and regulators. People who design and make nanomaterials need life-cycle tools that are not costly to develop and use, only require data inputs that a user could reasonably expect to have available, and are "fit for purpose" in the sense that they meet the needs of the user and the particular nanomaterial being evaluated. If nanomaterial designers and producers do not have the tools to apply life cycling thinking into their nanomaterial development decisions, then regulators will have greater difficulty in applying life-cycle thinking to regulatory evaluation.

To support building life-cycle assessment models, we need to consider what information is needed to make sustainability related decisions about nanomaterials and introducing them into and the environment. This would necessitate building a database with information on many types of nanomaterials.

The EPA has reviewed over 160 nanomaterials as new chemicals under the Toxic Substance Control Act (TSCA); these are materials not derived from chemicals already in the TSCA chemical inventory. The agency currently does not require nanomaterials derived from substances on the TSCA inventory to be submitted to the agency for review. Because of this, EPA likely is missing information on nanomaterials in commerce. Among other uses, such information would help EPA build out chemical categories and structural analogs for nanomaterials, which would facilitate review of reviewing new nanomaterials from life-cycle and sustainability perspectives.

Reporting Rule

The agency has issued a draft nanomaterial reporting rule under TSCA section 8(a). The proposal is for one-time reporting of nanomaterial substances derived from chemicals already on the TSCA inventory.

If a substance used to derive a nanomaterial is already on the chemical inventory, and it has been produced within the last three years, and has not been submitted to the agency before, the proposed reporting rule would require that existing information on that material be submitted to EPA for one-time reporting. Submitted information would include physical-chemical properties, material characterization, and environmental health and safety information.

The information would be reviewed to see if further analysis needs to be done. If the agency has questions about the nanomaterial, there may be the need for further discussion with the submitter.

This is not meant to create a public registry of nanomaterial substances. In addition, submitters have the right to claim their information as confidential.

This information is needed because there are gaps in knowledge about nanomaterials and what is being commercialized, since the EPA is seeing at this time only those nanoscale materials derived from substances that are not on the TSCA inventory.

The EPA has regulated new nanomaterial substances for about a decade, and we have no indication that there has been stigmatization of those materials. The agency believes that its pre-market review of new nanomaterials enhances public confidence in nanomaterials by using the best available science to evaluate nanomaterials' hazard and exposure potential throughout their life cycles prior ot approving them for commercialization.

The agency has proposed to demarcate reporting by focusing on substances in the range of 1-100 nanometers in at least one dimension. The proposed rule describes different environmental behavior-driving properties to determine whether a nanomaterial should be reportable, such as zeta potential, surface reactivity, and solubility. Generally, if there are two size ranges (in nanometers) derived from the same substance, and they are very different in those properties, experience suggests that they may behave differently in the environment and therefore may warrant separate reporting.

For new materials derived from chemicals on the TSCA inventory, the agency has asked that reporting be made within 135 days before a manufacturer's intent to commercialize. 135 days was chosen because the agency is required to review materials within 90 days, and experience has shown that 80% of the reviewed substances go into commercialization more than 45 days after that 90-day review.

The proposed rule was out for a 3-month comment period. The agency held a public meeting with web access to facilitate comment on various aspects of the rule. The EPA hopes to issue a final rule that is useful in enhancing public confidence, ensuring nanomaterials' safety, and provides information on nanomaterials to support their safe and environmentally sustainabile commercialization. More information is available in the references (*1*, *2*).

References

1. Morris, J.; Willis, J.; De Martinis, D.; Hansen, B.; Laursen, H.; Sintes, J. R.; Kearns, P.; Gonzalez, M. Science policy considerations for responsible nanotechnology decisions. *Nat. Nanotechnol.* **2011**, *6*, 73–77.
2. Nel, A.; Grainger, D.; Alvarez, P.; Badesha, S.; Castranova, V.; Ferrari, M.; Godwin, H.; Grodzinski, P.; Morris, J.; Savage, N.; Scott, N.; Wiesner, M. Nanotechnology Environmental, Health, and Safety Issues. In *Nanotechnology Research Directions for Societal Needs in 2020*; Roco, M. C., Mirkin, C. A., Hersam, M. C., Eds.; Springer: New York, 2011; pp 159-220, DOI: 10.1007/978-94-007-1168-6_5.

Chapter 10

Nanomaterials: Promise in Balance with Safety

Martin A. Philbert*

Dean and Professor of Toxicology,
School of Public Health, University of Michigan,
Ann Arbor, Michigan 48109, United States
***E-mail: philbert@umich.edu.**

Many nanomaterials have been or are in development for medical applications. Their potential benefits are great, particularly in view of the versatility and novel/emergent properties of nanomaterials. However, care needs to be exercised in terms of their safety. This article provides a brief review of some of the nanomaterials being developed for nanomedicine and medical devices, some of the current and next-generation products, and other related approaches. The safety aspects of nanomedicine are also emphasized.

Biomedical Uses and Applications

Nanobiomedicine boils down to some straightforward principles of physics and chemistry and the way that they interface with biology. Nanomaterials (nanoparticles) for biomedical applications are exploited for their enhanced biophysical properties. These properties include color, semiconduction, adhesion, storage, and enhancement of contrast in multiple domains (for MRI, ultrasound, optical, and other applications).

A recent analysis by NanoWerk (*1*) in Germany has identified all of the currently approved and investigational nanomaterials for commercial use as nanomedicines and devices. In the investigational area, there are 147 drugs and devices under development. Commercially, there are approximately 100 therapeutics and devices that have either crossed the IND phase and are in clinical trials or have been approved for use by the FDA.

Nanomedical Drugs and Devices

Among the drugs that are already approved or are in various stages of clinical trial and summarized in a recent report by International Federation of Pharmaceutical Manufacturers & Associations (*2*) is PEGylated doxorubicin, which is highly hydrophobic (*3*). It intercalates base pairs in DNA, and stabilizes DNA in rapidly dividing cells, causing the cancer to decrease its rate of growth. Within living memory, pharmacists would mechanically formulate such hydrophobic suspensions, and the resulting i.v. would be administered quickly before the emulsion 'broke.' The resulting infusion contained micelles of varying size resulting in complex pharmacokinetics, pain and frequently hemolysis. Using newer techniques in colloidal chemistry at the nanometer scale, the formulator has greater control of size and can generate monodisperse suspensions. The establishment of a stable colloid allows for the self-assembly of a number of components that can shield the active moiety (drug) from off-target sites and thereby improves efficacy while avoiding systemic toxicity. This principle has been used effectively in the manufacture of a number of nanmedicines targeted against cancer, e.g., Abraxane, and Daunorubicin -- used for the treatment of Kaposi's sarcoma and other cancers (*4*).

In 1994, the first nano-PEGylated enzyme to be approved for clinical use was L-asparaginase. It was developed by Enzon and used for acute lymphocytic leukemia. PEGylation can enhance the retention time of drug molecules and nanoparticles by protecting them from the immune system and degradation by circulating enzymes (*5*, *6*).

The newer combinations and modalities are moving away from simple hiding of a drug inside a shell and delivering it in a targeted or untargeted fashion to things like encapsulating gold nanoparticle in a hydrogel, and allowing a biologic tag such as Tumor Necrosis Factor to be used as a targeting and activating ligand. This can be activated with infrared or other forms of energy and is targeted for the superficial tumor of the head and neck. Similarly, gold nanoshells of various shapes and sizes allow for intravenous delivery with preferential accumulation in the tumor. The function of these gold nanoshells can be significantly enhanced by the conjugation of cancer drugs such as Doxil. Irradiation with infrared light within the biological window induces local heating of the surrounding tissue/cancer (*7*).

Metal and metalloid quantum dots (QD) are used with regularity for the tagging of cellular components or anatomical structures for basic research, and for the labeling and landmarking of cancerous tissues. The next generation of graphene quantum dots (GQD) allow for rational design and synthesis that imparts specific molecular, and hence biological properties for the detection and pH-dependent tagging of cancer cells (*8*). GQDs may exist as single sheets or may self-assemble into stacks depending on the electrolyte composition of the biological milieu. The resulting graphene composite may be conjugated with drugs or targeting moieties that can be induced to emit tuned emission wavelengths for easy detection. These materials are generally stable in a biological environment if made water-soluble. They may be PEGylated or hydroxylated to maintain their functional capability and emit light of the appropriate wavelength for detection (*9–11*).

Weissleder and his team at Harvard University (*12*) have developed a variety of magnetic and superparamagnetic iron oxide and Gd-containing nanoparticles that have been formulated in a variety of biocompatible matrices for infusion into the vascular space. These materials allow for the delineation of vascular tumors using magnetic resonance imaging and have been labeled with a variety of ligands that provide for specific accumulation in a variety of tumor types (*13*, *14*).

First Generation *Drug* Nanomedicines

The first generation of nanomedicines/drugs, those that are approved or in clinical trial, are generally small hydrophobic molecules, chemotherapeutics, antibiotics and antimycotics that are stabilized in a suitable matrix that makes the entire drug-containing entity more miscible with the electrolyte-rich environment of the plasma. Modification of the nanomaterial surface and the degree of crosslinking of the polymer chains or the composition of lipids in the micelle has effectively slowed fine-tuning of release of the active pharmaceutical ingredient and hence the pharmacokinetics/pharmacodynamics of the product.

On the horizon is the advent of vaccines attached to nanomaterials such as dendrimers that act as more effective adjuvants, thereby enhancing the potency of the elicited immunizing response. However, the rapid development of a bewildering array of potentially biocompatible nanomaterials provides for a greatly expanded toolbox of excipients, devices and combination products that will require new modes of testing for safety and efficacy.

Next Generation of Nanomedicines

Isacoff at the University of California-Berkeley is using protein and synthetic combinations of materials that, when inserted into a membrane, will permit light-controlled fluxes of calcium or potassium in membranes depending on the composition of the photo-switch/gate and associated ion channel. The photoswitches may be tuned to respond to infrared or ultraviolet light and have been used in the photoactive targeted labeling of neurons (*15*). Their use has also been proposed in the treatment of retinal blindness (*16*, *17*).

In a different mode of action, Bao and his colleagues at Georgia Tech (*18*) and others (*19*) are using the enormous forces generated by chaperonins in the cell for refolding misfolded proteins, such as tau, Aβ, APP, a-synuclein, and other proteins that appear in neurodegenerative diseases.

Sangamo Biosciences (*20*) in San Francisco has performed gene editing by modifying a series of zinc-finger proteins (ZFP) that are now in phase II clinical trials for the treatment of HIV/AIDS and earlier stage development that will address hemoglobinopathies (sickle cell and β-thalassemia), Hurler's and Hunter's syndromes, Huntington's Disease, lysosomal storage diseases and other monogenic diseases. These ZFPs may be elegantly targeted to produce double strand breaks at a precise location in the genome. Careful mutation of that locus may add or remove function from a receptor, enzyme of other functional protein.

Of particular relevance to those nanomaterials that have been designed to tag or treat tumors and cancers, enhanced permeability and retention within the body of the target tissue is effected by taking advantage of fenestrations in the blood vessels contained within the diseased tissue. Not every blood vessel in the body is completely 'tight'. Most blood vessels are leaky, having pores in the range of 6 – 12 nm in tissues such as skin and as large as ~ 5μm in the terminal capillaries of the red pulp in the spleen. In the liver, vascular pores are of the order of 180 nm in diameter. These fenestrations provide an efficient mechanism for bulk macromolecular nutrients out of the blood vessels. Therefore, the use of leaky blood vessels in highly vascular cancers with poorly formed, more embryonal vasculature brings with it the potential for off-target toxicity as the nanoparticles may accumulate in the liver and other tissues with porous blood vessels. The 'object permanence' of nanoparticles and the general lack of distribution of the components of more durable nanomaterials with the fat or water compartments of the body may cause the deleterious build up of individual nanoparticles which may agglomerate in place and become entrapped in extra-/intracellular complexes (with or without calcium) that remain in place in the tissue. The native ability of the now immobilized nanomaterial to invoke numerous rounds of inflammation in an attempt to dispose of the submicron – micron-sized complexes may form the substrate for oxidative damage to key cellular macromolecules such as DNA, cell membranes, and/or signaling proteins that regulate cell death and/or cell death.

Using ^{14}C-radiolabeled polymer nanoparticles (polyacrylamide hydrogel ~ 35 nm hydrodynamic radius), our team has been able to trace the major portion of an intravenous dose in the liver. However, closer microscopic inspection of the tissue reveals that intact material readily agglomerates and is actively taken up by the Küpffer cells that line the surfaces of liver sinusoids (a highly adapted capillary space). Küpffer cells are a critical component of the reticuloendothelial system that protects the internal compartments of the body from the deleterious effects of particulate matter that escapes the barriers of the gut, skin or other epithelial boundaries that keep infectious agents and particulate matter out. These cells have evolved to recognize a number of biological and physical materials in the nanometer and sub-micrometer range and are rimed to dispose of theme through the use of inflammatory cytokines, oxidizing enzyme systems and low pH. Therefore, an inability to clear deposited nanomaterials from the extracellular spaces will inevitably result in numerous rounds of inflammation until the surface is sufficiently passivated or walled off by scar tissue.

Synthesis and Manufacture of Nanomaterials

There exist a wide range of manufacturing processes that produce nanomaterials with specific bulk and surface properties of the finished product. Milling of bulk materials to the required diameter remains the simplest process, however, other processes such as thermal, crystalline, aerosols, sonic and/or chemical cavitation, colloidal suspensions, etc., for the production of sol/gels, polymer matrices, and mineral/metal or carbonaceous nanomaterials have been successfully deployed. Even though the chemical identity, size and physical properties of the nanomaterial may be essentially the same, the toxicologic potential of the finished product is significantly affected by the mode of synthesis and the presence of catalysts, unwanted synthetic by-products (which may or may not affect the final performance and biological activity of the synthesized nanomaterial). Moreover, lingering trace contaminants of synthesis that would not ordinarily accumulate in the body, absent a particle that retains, and accumulates, the toxicant in a specific anatomical location may become of toxicologic concern. The high surface area-to-volume ratio in nanomaterials makes them incredibly useful. There is excellent potential for adsorption of unwanted synthetic materials on the massively increased surface and enhanced possibilities for nuisance surface chemistries that alter biological/ecological macromolecules. For labile nanoparticles, there is the possibility of potential release of contaminants into biological environments. The rapid increase in the combinations and permutations of chemistries that may be readily used to produce a wide array of nanomaterials in large quantity and with few barriers to synthesis will far outstrip our capacity to proactively assess the toxicity/safety of each potential formulation. For these, and other reasons, the manufacture of nanomaterials for biomedical and other applications needs to be carefully designed for safety and should incorporate the principles of green chemistry (*21*).

Safety Considerations

Rules for Safety of Nanomaterials

The Rule of Five (*22*) was originally formulated by Lipinski in 1997 and has been successfully used to predict the safety/toxicity of low molecular weight therapeutics developed for oral administration. The rule describes those physical and chemical properties that are major factors in the determination of whether or not a drug will accumulate in the body and/or produce deleterious effects. The major predictors of safety are that a compound breaks no more than one of the following: MW ≤ 500, octane-water partition coefficient log $P \leq 5$, H-bond donors ≤ 5, H-bond acceptors (sum of N and O atoms) ≤ 10, a polar surface area of ≤ 140 Å^2, or sum of H-bond donors and acceptors ≤ 12, and rotatable bonds ≤ 10. Drugs with these characteristics have a good chance of making through the FDA. The majority of nanomaterials in existence violate at least 2 or 3 of these rules. However, Scott McNeil and his colleagues at the National Cancer Institute's Nanotechnology Characterization Laboratories (*23*) have completed physical, chemical and biological tests of safety on a wide range of nanomaterials intended for clinical use. They have uncovered a series of factors that render nano-formulations safe for use (*24*). They have also provided useful advice on nanoformulation and on selection of nanotechnology carriers.

It took the better part of a century for us to understand the properties of small molecules that predict safety/toxicity and studies of the toxicity of biologics and biosimilars is an ongoing process. However, as the field of nanomedicine matures, we will need to be vigilant and careful in the design and deployment of this useful class of novel materials.

References

1. Nanowerk website provides educational information on nanotechnology. The homepage is at http://www.nanowerk.com (accessed 1/28/2016).
2. International Federation of Pharmaceutical Manufacturers & Associations; http://www.ifpma.org/fileadmin/content/Publication/2014/IFPMA_-_Facts_And_Figures_2014.pdf (accessed 10/31/15)
3. Gabizon, A. A. Pegylated liposomal doxorubicin: metamorphosis of an old drug into a new form of chemotherapy. *Cancer Invest.* **2001**, *19*, 424.
4. Ojima, I.; Lichtenthal, B.; Lee, S.; Wang, C.; Wang, X. Taxane anticancer agents: a patent perspective. *Expert Opin. Ther. Pat.* **2016**, *26*, 1–20.
5. Mu, X.; Qiao, J.; Qi, L.; Ma, H. (2015) Poly(2-vinyl-4,4-dimethylazlactone)-functionalized magnetic nanoparticles as carriers for enzyme immobilization and its application. *ACS Appl. Mater. Interfaces* **2015**, *6*, 21346–21354.
6. Bahreini, E.; Aghaiypour, K.; Abbasalipourkabir, R.; Mokarram, A. R.; Goodarzi, M. T.; Saidijam, M. Preparation and nanoencapsulation of l-asparaginase II in chitosan-tripolyphosphate nanoparticles and in vitro release study. *Nanoscale Res. Lett.* **2014**, *9*, 340.
7. Noh, M. S.; Lee, S.; Kang, H.; Yang, J. K.; Lee, H.; Hwang, D.; Lee, J. W.; Jeong, S.; Jang, Y.; Jun, B. H.; Jeong, D. H.; Kim, S. K.; Lee, Y. S.; Cho, M. H. Target-specific near-IR induced drug release and photothermal therapy with accumulated Au/Ag hollow nanoshells on pulmonary cancer cell membranes. *Biomaterials* **2015**, *45*, 81–92.
8. Qiu, J.; Zhang, R.; Li, J.; Sang, Y.; Tang, W.; Rivera, G. P.; Liu, H. Fluorescent graphene quantum dots as traceable, pH-sensitive drug delivery systems. *Int. J. Nanomedicine* **2015**, *28*, 6709–6724.
9. Mehra, N. K.; Jain, K.; Jain, N. K. Design of multifunctional nanocarriers for delivery of anti-cancer therapy. *Curr. Pharm. Des.* **2015**, *21*, 6157–6164.
10. Sao, R.; Vaish, R.; Sinha, N. Multifunctional drug delivery systems using inorganic nanomaterials: A Review. *J. Nanosci. Nanotechnol.* **2015**, *15*, 1960–1972.
11. Prabhakar, N.; Nareoja, T.; von Haartman, E.; Sen Karaman, D.; Burikov, S. A.; Dolenko, T. A.; Deguchi, T.; Mamaeva, V.; Hanninen, P. E.; Vlasov, I. I.; Shenderova, O. A.; Rosenholm, J. M. Functionalization of graphene oxide nanostructures improves photoluminescence and facilitates their use as optical probes in preclinical imaging. *Nanoscale* **2015**, *21*, 10410–10420.
12. Weissleder, R. https://csb.mgh.harvard.edu/weissleder (accessed 10/31/2015).
13. Kuijten, M. M.; Hannah Degeling, M.; Chen, J. W.; Wojtkiewicz, G.; Waterman, P.; Weissleder, R.; Azzi, J.; Nicolay, K.; Tannous, B. A. Multimodal targeted high relaxivity thermosensitive liposome for in vivo imaging. *Sci. Rep.* **2015**, *5*, 17220.
14. Miller, M. A.; Gadde, S.; Pfirschke, C.; Engblom, C.; Sprachman, M. S.; Kohler, R. H.; Laughney, A. M.; Wojtkiewicz, G.; Kamaly, N.; Bhonagiri, S.; Pittet, M. J.; Farokhzad, O. C.; Weissleder, R. Predicting therapeutic nanomedicine efficacy using a companion magnetic resonance imaging nanoparticle. *Sci. Transl. Med.* **2015**, *7*, 314–183.

15. Berlin, S.; Caroll, E. C.; Newman, Z. L.; Okada, H. O.; Quinn, C. M.; Kallman, B.; Rockwell, N. C.; Martin, S. S.; Lagarias, J. C.; Isacoff, E. Y. Photoactivatable genetically encoded calcium indicators for targeted neuronal imaging. *Nat. Methods* **2015**, *12*, 852–858.
16. Patty, R.; Isacoff, E. Y. Cooperative Binding of Stromal Interaction Molecule 1 (STIM1) to the N and C Termini of Calcium Release-activated Calcium Modulator 1 (Orai1). *J. Biol. Chem.* **2016**, *291*, 334–341.
17. Gaub, B. M.; Berry, M. H.; Holt, A. E.; Osacoff, E. Y.; Flannery, J. G. Optogenetic Vision Restoration Using Rhodopsin for Enhanced Sensitivity. *Mol. Ther.* **2015**, *23*, 1562–1571.
18. Bao, G. http://www.amplifier.gatech.edu/experts/gang-bao (accessed 10/31/2015).
19. Hoersch, D.; Roh, S. H.; Chiu, W.; Kortemme, T. Reprogramming an ATP-driven protein machine into a light-gated nanocage. *Nat. Nanotechnol.* **2013**, *8*, 928–932.
20. *Sangamo BioSciences*. http://www.sangamo.com/index.html (accessed 10/31/ 2015).
21. Albrecht, M. A.; Evans, C. W.; Raston, C. L. Green chemistry and the health implications of nanoparticles. *Green Chem.* **2006**, *8*, 417–432.
22. Lipinski, C. A.; Lombardo, F.; Dominy, B. W.; Feeney, P. J. Experimental and computational approaches to estimate solubility and permeability in drug discovery and development settings. *Adv. Drug Delivery Rev.* **2001**, *46*, 3–26.
23. National Cancer Institute website at ncl.cancer.gov (accessed 1/28/2016).
24. Dobrovolskaia, M. A.; McNeil, S. E. Strategy for selecting nanotechnology carriers to overcome immunological and hematological toxicities challenging clinical translation of nucleic acid-based therapeutics. *Expert Opin. Drug Delivery* **2015**, *12*, 1163–1175.

Chapter 11

Opportunities and Challenges for Health, Safety, and the Environment: The Regulatory Void?

Lynn L. Bergeson*

Managing Partner and Owner Bergeson & Campbell, P.C. Suite 100W, 2200 Pennsylvania Avenue, N.W. Washington, DC 20037, United States
***E-mail: lbergeson@lawbc.com.**

Governance paradigms are essential to the commercialization and sustained development of new technologies, including nanotechnologies. International governance bodies are evolving to address the benefits of nanotechnologies while seeking to manage their potential risks to human health and the environment through a variety of voluntary, standard-setting, regulatory, statutory, and related governance platforms. Whether and how each is working is very much a work in progress. What appears to be emerging, however, is a consensus that a variety of methods are needed, and regulatory measures are integral to the success of effective governance paradigms. This article will survey the governance approaches that are emerging, with particular emphasis on the need for regulatory measures in targeted areas to ensure the integrity of core governance principles and provide some measure of commercial predictability.

The designated topic on which I was asked to present was "Nanotechnology: Delivering on the Promise -- Opportunities and Challenges for Health, Safety, and the Environment: The Regulatory Void." I took some liberty in adding a question mark after "void" as it is entirely unclear whether a regulatory void exists and, if so, would it necessarily prevent nanotechnology from delivering on all that it is expected to provide. What is, after all, a "void" in this context?

As an adjective, void means, among other things, not valid or legally binding as in "to void a contract." When used as a noun, it means "a completely empty space," uninhabited, or unoccupied. The word "void" is, thus, hardly neutral, particularly in this context. It conjures up images and impressions that are unflattering.

With regard to the regulation of nanotechnology, none of these terms feels right. Yet the reverse seems equally ill-fitting. Upon reflection, my view is we, as a community of nanotechnology stakeholders, are getting there, the environmental, health, and safety space is not properly characterized as a void, but there are "voids" that need to be filled, and our challenge as a community of stakeholders is to determine how best to do so effectively and quickly.

As a relatively new technology, nanotechnology tests us at many levels. Scientists know that nanotechnology challenges us at the scientific level with regard to understanding nanoscale materials' chemical, toxicological, and biological properties, exposure pathways, and identifying potential opportunities where nanomaterials may pose risk. Commercially, any new technology tests even the most gifted entrepreneur. The road to commercialization is unpredictable, and statistically likely to result in more failures than successes. Most importantly for the legal community, nanotechnology tests the core integrity of existing governance tools -- laws, regulations, guidance, and everything else along the governance spectrum -- to identify and manage potential risks, as appropriate, without stifling nanotechnology's promise of a better tomorrow. This is not easy, as we all know.

As stewards of this technology, even a basic sense of history compels us to be mindful of prior mistakes, and encourages us to do things differently from another big "it" -- biotechnology. Debates to this day vascillate between images of a utopian world heralding a better tomorrow and a dystopian collapse of human civilization as we know it. This experience is instructive as to what we can and should do differently.

On the governance side, in managing emerging technologies we have adapted the rubber-suit approach -- the stretching, contorting, and adapting of existing legal and regulatory tools to address potential risk posed by emerging technologies. The very speed of innovation leaves us little choice. Existing authorizations bestowe upon federal agencies significant authority under various federal laws, the implementation of which must relentlessly "adapt" to manage potential risks occasioned by emerging technologies, including nanotechnology. Some believe that our legal infrastructure is sufficiently robust for this purpose. The American Bar Association (ABA) Section of Environment, Energy, and Resources so concluded ten years ago. As then Chair of the ABA Environment, Energy, and Resources Section, it was my honor to lead an effort to prepare a series of legal briefs on exactly this topic. Dozens of lawyers contributed to the effort and we carefully reviewed all primary federal environmental statutes, including the Clean Air Act, the Clean Water Act, the Toxic Substances Control Act (TSCA), and the Federal Insecticide, Fungicide, and Rodenticide Act (FIFRA), among others, and concluded they were sufficiently robust as written to manage risks from nanotechnology (*1*). Plainly, however, gaps, or "voids," existed then and exist today. These voids exist at many levels -- lack of resources, lack of data, and lack of participation from industry and other stakeholders for whatever reason to contribute information, among other gaps. Our challenge is to fill these voids as quickly as possible.

Similarly, new technologies test political will. One of our biggest challenges as a community, certainly for lawyers in this area, is navigating changes in Administrations, in career staff leadership positions, in agency priorities and policies, in diminishing resources, in addressing the potentially disruptive consequences of a growing exodus of senior agency personnel, and in differing levels of technological literacy in Congress and the agencies. These changes similarly create voids and pose formidable challenges in aligning program goals within and among federal agencies and expressing those goals coherently and consistently through the federal family.

With regard to the regulation on nanotechnology, there is no dirth of federal regulatory activity. Many agencies represented in the National Nanotechnology Initiative (NNI) are actively engaged in regulatory action of one form or other -- Food and Drug Adminstration (FDA), Consumer Product Safety Commission (CPSC), Environmental Protection Agency (EPA), Occupational Safety and Health Administration (OSHA), and National Institute for Occupational Safety and Health (NIOSH), most especially. NIOSH is mentioned here not because it regulates, but because its contributions to industry and other nano stakeholders with regard to the development of safe handling practices and related areas are so significant. I never pass on an opportunity to thank NIOSH Director John Howard's leadership and Chuck Geraci's and his colleagues' significant contributions to the field (*2*). Yes, there are gaps or voids in these regulatory efforts, but certainly the space is by no means empty.

EPA's regulatory contributions are also significant. There are voids in EPA's understanding of the toxicological properties and exposure characteristics of nanomaterials. This is because there is limited information available on these materials. EPA, industry, and other stakeholders are working to fill these gaps. It is important to do so as we must have a better understanding of the toxicological profiles and exposure implications of nanoscale materials to ensure we protect human health and the environment.

As a legal matter, some nanoscale chemical substances are considered "new" under TSCA and require notification to EPA as a predicate to marketing the nanoscale chemical (*3*). EPA's authority under TSCA is an important tool that adds to our knowledge of nanoscale materials and helps ensure that the commercialization of nanoscale materials pose no unreasonable risk. Over the past decade, EPA has received some 160 premanufacture notifications for new nanoscale substances. Most have been allowed into commerce, but with restrictions. One hundred percent undergo detailed review, which takes 6-24 months, considerably longer than the statutorily required 90 days under TSCA.

There is no such premarket notification requirement for nanoscale materials considered "existing" chemical substances under TSCA. This is because EPA's TSCA Inventory Policy, issued in 2008, provides that nanoscale versions of chemicals listed on the TSCA Inventory are themselves considered existing if the nanoscale version shares the same molecular identity as its conventional counterpart (*4*). This policy means that EPA's ability to compel the submission of information on nanoscale versions of existing chemical substances must be the subject of a federal rulemaking under TSCA.

Some of these "gaps" could be filled as EPA proposed recently a TSCA rule that would compel the submission of information to EPA under TSCA Section 8(a) (*5*). The proposal has been many years in the making and illustrates a lack of alignment of priorities, common goals, and regulatory objectives between EPA, the Office of Management and Budget (OMB), and other federal agencies. EPA has proposed one-time reporting and recordkeeping requirements for certain chemical substances already in commerce when they are manufactured or processed as nanoscale materials. Specifically, EPA proposed requiring companies that manufacture or process (or intend to manufacture or process) chemical substances in the nanoscale range to report electronically information, including the specific chemical identity, production volume, methods of manufacture, processing, use, exposure and release information, and available health and safety data. The proposed rule would apply to chemical substances that have unique properties related to their size. The proposed rule provides exclusions for chemical substances in the nanoscale range that would not be subject to the rule. The rule, when issued in final, will be an important tool to help fill some significant information voids.

Moving to EPA's pesticide office, the Office of Pesticide Programs (OPP) issued a proposed policy on nanoscale materials in pesticides in 2011. OPP proposed obtaining information on such materials by using either its "adverse effects" reporting mechanism, or its data call-in authority under FIFRA. Unsurprisingly, the adverse effects approach met with significant push-back. That policy was issued four years ago and is still in a state of suspended animation. This has created a directional void that is overdue for clarification. Importantly, EPA has registered a nanopesticide -- nanosilver in textile applications -- and was sued almost immediately for it (*6*). There are clear gaps in OPP's knowledge and practice in registering nanopesticides. This is one area where industry needs to step up, and where OPP needs to reformulate its outdated and largely "voided" 2011 policy position.

EPA's Water Office has also expressed interest in nanoscale materials. In September 2014, EPA announced its decision to collect information on wastewater discharge hazards associated with nanomaterials manufacturing and processing (*7*). Specifically, EPA stated that it is collecting data and information on the potential industrial wastewater discharge hazards associated with nanomaterials manufacturing and formulating. EPA requested public comment and stakeholder input relating to any information or data available on the wastewater hazards and discharges associated with the manufacture of nanomaterials and their use in manufacturing or formulating products, as well as any other information believed to be relevant. This information will assist in addressing voids in EPA's Office of Water's understanding of nanoscale materials in water discharges.

EPA is engaged at many different levels in the governance of nanoscale materials, including at the international level. Its long engagement with the Organisation for Economic Cooperation and Development (OECD) and its Working Party on Manufactured Nanomaterials has yielded significant contributions to our knowledge and understanding of the properties and risks of nanoscale materials. EPA's work with Canada under the Regulatory Cooperation Council (RCC) is equally beneficial. Of course, EPA's development of test methods and data to address hazards and exposures throughout the OECD, EPA's own Office of Research and Development, and collaborations with federal partners continues to fill important knowledge voids. Collectively, these initiatives will better integrate data into risk assessment and risk management decision-making (*8*).

This brings me to my concluding thoughts. Federal policy is and has been supportive of delivering nanotechnology's fundamental promise. The federal government's approach to delivering on that promise through an integrated, coherent, efficient, and effective regulatory approach across federal agencies is a work in progress. Voids plainly exist and need to be filled to diminish incoherence, inconsistency, and directional uncertainty, all of which frustrate commercial development and invite disarray in business. Significant private sector support is needed to ensure the private sector is properly engaged; enhanced communication and collaboration among stakeholders is needed to leverage effort; and strong federal leadership is needed to ensure alignment on core policy objectives and goals.

References

1. The Section prepared issue papers on each federal environmental statute, and one on expanding governance systems. The papers were later published in a book entitled *Nanotechnology Environmental Law, Policy, and Business Considerations*; Bergeson, L. L., Ed.; ABA Press, 2010.
2. Centers for Disease Control and Prevention. *The National Institue for Occupational Safety and Health (NIOSH)*; http://www.cdc.gov/niosh/ (accessed 5/8/2015).
3. Bergeson, L.; Plamondon, J. TSCA and Engineered Nanoscale Substances. *Nanotechnology Law & Business;* **2009** March.
4. *TSCA Inventory Status of Nanoscale Substances General Approach*; http://www2.epa.gov/tsca-inventory/nanoscale-substances-tsca-inventory (June 23, 2008).
5. Federal Register. Vol 80, pp 18330 (Apr. 6, 2015); https://www.gpo.gov/fdsys/pkg/FR-2015-04-06/pdf/2015-07497.pdf.
6. On November 7, 2013, the U.S. Court of Appeals for the Ninth Circuit granted in part and denied in part the Natural Resources Defense Council's (NRDC) petition for review of the EPA's conditional registration of HeiQ AGS-20 and AGS-20 U (collectively, AGS-20) nanosilver products. See http://cdn.ca9.uscourts.gov/datastore/opinions/2013/11/07/12-70268.pdf. The court held that "substantial evidence" supported EPA's decision to use the characteristics of toddlers rather than infants in determining whether AGS-20 placed consumers at risk. The court vacated EPA's decision "insofar as it concluded that there was no risk concern requiring mitigation for short- and intermediate-term aggregate oral and dermal exposure to textiles that are surface-coated with AGS-20." The court stated: "Our decision to

grant the petition and vacate the decision in part is based solely on the fact that EPA's own rule states that there is a risk concern requiring mitigation when the calculated MOE is less than or equal to 1,000 and, under these circumstances, the actual MOE equals 1,000. This holding does not affect any portion of EPA's decision where the calculated MOE is greater than 1,000." Finally, the court held that "substantial evidence supported the EPA's decision not to consider additional sources of exposure to nanosilver other than AGS-20 in concluding that the product would not have adverse effects on consumers."

7. Federal Register. Vol 79, pp 55472 (Sept. 16, 2014); https://www.gpo.gov/fdsys/pkg/FR-2014-09-16/pdf/2014-22062.pdf.
8. United States Environmental Protection Agency (EPA). *Research on Nanomaterials*; http://www2.epa.gov/chemical-research/research-evaluating-nanomaterials-chemical-safety (accessed 5/8/2015).

Chapter 12

Adaptive Governance for the Nanotechnology Workplace

Vladimir Murashov* and John Howard

National Institute for Occupational Safety and Health,
395 E. Street, SW, Suite 9200, Washington, DC 20201, United States
***E-mail: vem8@cdc.gov.**

Adaptive governance is commonly used in reference to the responsible introduction of emerging technologies such as nanotechnology into the society. A component of adaptive governance for nanotechnology, the proactive risk mitigation for nanotechnology workplaces, is presented in this paper.

Introduction

The notion of adaptive governance emerged about two decades ago from the intersection of two areas of inquiry the application of ecological systems theory to natural resource management and the study of self-governing institutions (*1*).

There are two components to the term: governance and adaptation. Governance refers to "the institutional arrangements which shape actors' decisions and behavior within groups or organizations (such as firms or nations)" (*1*). Governance is in essence a fluid mediating the formal and informal negotiations of decisions between actors with different forms and degrees of influence. Adaptation refers to an unmanaged, but systematic process of change in response to competitive pressures. Thus, adaptive governance refers to the evolution of the rules and norms that promote the realization of underlying human needs and preferences given changes in understanding, objectives, and the social, economic and envrionmental context (*1*). It operates at the high societal level. The component of adaptive governance for nanotechnology which is of interest to the U.S. National Institute for Occupational Safety and Health is the proactive or anticipatory risk mitigation for nanotechnology workplaces.

Proactive/Anticipatory Risk Mitigation for Nanotechnology Workplaces

Proactive risk mitigation for an emerging technology such as nanotechnology represents a number of challenges, opportunities and solutions (*2*). The main challenge is deciding how to best incorporate higher levels of uncertainty in assessing risks and controls into the standards development process (*3*).

The primary opportunities presented by nanotechnology include 1) the possibility to evaluate and mitigate risks of nanomaterials throughout the life cycle stages of nanomaterials; and 2) the opportunity to design hazards out and to minimize exposures through material and process design. In the life-cycle approach nanomaterial safety is evaluated through stages of nanomaterial life starting from the production of raw nanomaterials to manufacturing of consumer products, to the use of products containing nanomaterials, to the product end of life when it goes to landfills, incinerated or recycled or ends up in the environment through other wastestreams (*4*). Worker exposure to nanomaterials can occur at any of these stages. The life-cycle approach was recognized by the U.S. National Nanotechnology Initiative (NNI) as one of research needs in safety research: "Application of adaptive management tool to evaluate life cycle analysis implementation" (*4*). Thus, the NNI strategy for environmental, health and safety of nanotechnology also highlighted the application of the adaptive approach to nanomaterials safety.

The next opportunity in risk management of nanotechnology is the opportunity to design hazards and exposures out through nanomaterial and process design. In the workplace, safety is managed through the so-called hierarchy of controls (*5*). In this approach the first step is to attempt elimination of hazardous nanomaterials or replacement with less hazardous alternatives. If this first step does not bring the risk to an acceptable level, the next step is to reduce exposure through the use engineering controls. If after these steps, the risk remains high, administrative controls such as training and job assignment rotations, are implemented. Finally, personal protective equipment, such as respirators and gloves, is recommended as the last line of defense.

Due to the great flexibility of nanomaterials in their size, shape, chemical composition resulting in a wide spectrum of possible physico-chemical properties, nanomaterials are particularly well-suited for the application of the proactive risk mitigation at the first step of the hierarchy of controls: hazard elimination or substitution with a less hazardous analog. For example, hazards of nanomaterials could be eliminated or reduced by avoiding hazardous elements such as regulated heavy metals in their chemical composition, improving their biodegradability, utilizing safer formulations by avoiding toxic solvents and improving safety of manufacturing processes.

The primary solutions for proactive risk mitigation of nanomaterials in the workplace include employing 1) prudent measures to mitigate exposures when a paucity of hazard data exists; 2) qualitative anticipatory risk mitigation tools; 3) integration of soft (voluntary) and hard law (mandatory) approaches; 4) broad expert and stakeholder participation through public-private partnerships; and 5) sharing best available mitigation strategies among stakeholders.

Prudent Measures To Mitigate Exposures

Several calls have been made to implement prudent measures to mitigate exposures to potentially hazardous nanomaterials (*6, 7*). This can be accomplished with conventional engineering controls, such as local exhaust ventilation, for mitigating exposure to airborne materials. It has been shown that conventional controls are effective at reducing the amount of nanomaterials in workplace air when chosen wisely and employed properly (*8*).

Qualitative Anticipatory Risk Mitigation Tools

An example of a qualitative anticipatory risk mitigation tool is control banding for nanomaterials developed by ISO (*9*). In this proactive approach hazard potential of nanomaterials is assessed using available information about toxicity and physico-chemical properties. As a result of this analysis, nanomaterials are assigned into one of five hazard bands. Similarly, exposure potential to nanomaterial in the workplace is assessed using available information about the form of nanomaterial (powder, liquid dispersion, solid dispersion in polymer matrix, etc.) and the type of manufacturing process (mechanical reduction, wet chemistry, gas phase synthesis, etc.). Based on that assessment, each exposure situation is assigned into an exposure potential band. A combination of the hazard band and the exposure potential band would produce a recommended control option out of five possible ranging from using general ventilation to full containment and seeking expert advice.

Qualitative exposure assessment is also a part of a harmonized tiered approach for nanomaterial exposure assessment in the workplace recently published by the Organization for Economic Cooperation and Development (OECD) (*10*). In this approach, the first tier starts with information gathering to evaluate whether any exposure to nanomaterials is possible. If the answer is "yes," the second tier, basic exposure assessment is conducted using readily available and inexpensive devices such as particle counters. If these measurements indicate that the concentration of airborne particles is significantly increased over the background levels and their origin is known, then additional risk management measures are implemented to reduce the exposure. Otherwise the third tier, expert exposure assessment is conducted. It includes the use of sampling equipment and subsequent chemical analysis.

Integration of Soft and Hard Law Approaches

In the proactive risk mitigation there is a paucity of data to develop hard regulations. Therefore, in this regime it is necessary to operate at the bottom of the regulatory pyramid, which relies on information gathering and multi-stakeholder norms and self-regulation (*11*). In this regime international voluntary occupational safety and health standards play an increasing role in shaping national regulatory standards. International standards can be 1) nationalized; 2) used to support enforcement of the employer's general duty to provide work free of recognized hazards; and 3) adopted as voluntary guidance by the government (*12*).

There are a number of international standards developing organizations active in the area of nanomaterial safety. The most active and influential are the International Organization for Standardization (ISO) and the Organization for Economic Cooperation and Development (OECD). The ISO Technical Committee 229 (TC229) Nanotechnologies has four working groups (*13*). WG1 on terminology is presently working on core terms for nanotechnology and nanomaterials, terminology for two-dimentional nanomaterials, nanocellulose and quantum phenomena. WG2 on measurements is developing standards for characterization of nanocellulose and graphene, for using UV-Vis absorption to characterize cadmium chalcogenide, mass-spectroscopy to characterize single nanoparticles, and electron microscopy to measure size of nanoparticles. WG3 on health and safety is developing standards on *in vitro* toxicity testing of nanomaterials, a framework for setting occupational exposure limits for nanomaterials, and characterization of nanomaterials for their risk assessment. WG4 on material specification is developing standards for carbon nanotube dispersions. As of July 22, 2015 this technical committee published a total of 43 standards of which 13 were prepared by WG3 and deal directly with the safety and health issues of nanomaterials (*14*).

OECD Working Party on Manufactured Nanomaterials has four steering groups. Steering group on testing and assessment is in the process of publishing dossiers with data on toxicity testing and physico-chemical characterization for eleven nanomaterials. It is also responsible for updating test guidelines to make them suitable for nanomaterials. Steering group on risk assessment and regulatory programs is looking at interspecies variability factors in toxicity studies and at dissolution as a function of surface chemistry. Steering group on exposure measurement and mitigation is developing a protocol for measuring carbon nanotubes in the air and a report on measuring biodurability of nanomaterials. Finally the fourth steering group is looking at environmentally sustainable use nanomaterials. As of July 22, 2015 this working party published 57 reports (*15*).

Public-Private Partnerships

In order to develop proactive safety standards, public-private partneships are critical. In the workplace partnerships between government research organizations, nanotechnology manufacturers and downstream users, workers, academic researchers and safety and health practitioners are needed to collaboratively develop risk assessment and risk control strategies to eliminate worker risk and help achieve nanotechnology's promise (*16*). Aims of such public-private partnerhisps would be 1) protecting workers by encouraging implementation of prudent exposure mitigation measures; 2) promoting nanotechnology risk assessment and risk mitigation research; 3) collecting and sharing exposure information among nanotechnology workplaces; 4) identifying and studying the use of candidate occupational risk mitigation practices; and 5) developing the evidence base to provide protection for workers (*16*). An example of a public-private partnership for nanotechnology workplace safety is the NIOSH field team effort (*17*). Since mid-2006 when this team was established, it conducted on a volutary basis over 100 visits to 65 different sites.

The team investigated potential exposures in a diversity of sites, nanomaterials and applications using knowledge and experience gained to advance guidance and recommendations for ensuring safety of workers in nanotechnology workplaces. Partnerships with the private sector is a key to the success of the NIOSH field team.

There are other examples of NIOSH partnerhsips with the private sector. In 2014 NIOSH signed a memorandum of understanding with the College of Nanoscale Science and Engineering at SUNY Polytech Institute in Albany, NY to advance research and guidance on occupational safety and health of nanoelectronics (*18*). NIOSH guidance on safe practices for working in research laboratories (*19*) was developed under a memorandum of understanding with the Center for High-Rate Nanomanufacturing at Northeastern university. Formal NIOSH collaborations with the Center for Multifunctional Polymer Nanomaterials and Devices has focused on comprehensive evaluation of manufacturing processes, assessing worker exposures, evaluating exposure control methods and making recommendations for improvement, incorporating safe and sustainable design in facilities and processes, and providing an effective risk management framework suitable for small start-up businesses. NIOSH collaboration with the Center for Biological and Environmental Nanotechnology and International Council on Nanotechnology at Rice University produced GoodNanoGuide web-based guidance and training materials (*20*).

Coordination

With so many players actively involved in ensuring the safety of nanotechnology, close coordination is necessary in order to optimize the use of limited resouces. Coordination can be realized at several levels: 1) coordination of research, e.g. US-EU Communities of Research on nanoEHS; 2) coordination of regulatory programs (*e.g.* Canada-US Regulatory Cooperation Council); 3) coordination of standards development among government-level international organizations (*e.g.* United Nations, OECD), private international standards developing organizations (*e.g.* ISO, ASTM International) and multi-stakeholder *ad hoc* organizations (*e.g.* International Alliance for NanoEHS Harmonization, Global Measurement Harmonization Workgroup, NanoRelease).

Concluding Remarks

In conclusion, proactive risk mitigation of nanotechnology in the workplace follows a well established framework. In this framework risk is first anticipated, then recognized and evaluated. Appropriate levels of controls are applied and their effectiveness is confirmed. This cycle is repeated as necessary to reduce risk to an acceptable level and to capture on-going changes in the workplace. Only through partnerships among industrial hygienists, toxicologists and other researchers, engineers, businesses, and regulators, can this framework be applied proactively to emerging tehcnologies such as nanotechnology. By working together we can be successful in the safe introduction of nanotechnology into our everyday lives.

Disclaimer: The findings and conclusions in this report are those of the authors and do not necessarily represent the views of the National Institute for Occupational Safety and Health.

References

1. Hatfield-Dodds, S.; Nelson, R.; Cook, D. C. Adaptive governance: an introduction, and implications for public policy. *ANZSEE Conference* **2007**, 1–13.
2. Murashov, V.; Howard, J. Essential features for proactive risk management. *Nat. Nanotechnol.* **2009**, *4*, 467–470.
3. Murashov, V.; Howard, J. The US must help set international standardsfor nanotehcnology. *Nat. Nanotechnol.* **2008**, *3*, 635–636.
4. National Science and Technology Council. *National Nanotehcnology Initiative: Environmental, Health, and Safety Research Strategy*; October 2011.
5. National Institute for Occupational Safety and Health. *Hierarchy of controls.Workplace Safety & Health Topics*; available at http://www.cdc.gov/niosh/topics/hierarchy/ (accessed on October 5, 2015).
6. Schulte, P. S.; Geraci, C. I.; Murashov, V.; Kuempel, E. D.; Zumwalde, R. D.; Castranova, V.; Hoover, M. D.; Hodson, L.; Martinez, K. F. Occupational safety and health criteria for responsible development of nanotechnology. *J. Nanopart. Res.* **2014**, *16*, 2153.
7. National Institute for Occupational Safety and Health. *Approaches to safe nanotechnology: managing the health and safety concerns associated with engineered nanomaterials*; DHHS (NIOSH) Publication No. 2009−125; U.S. Department of Health and Human Services, Centers for Disease Control and Prevention, National Institute for Occupational Safety and Health: Washington, DC, 2009
8. National Institute for Occupational Safety and Health. *Current strategies for engineering controls in nanomaterial production and downstream handling processes*; DHHS (NIOSH) Publication No. 2014–102; U.S. Department of Health and Human Services, Centers for Disease Control and Prevention, National Institute for Occupational Safety and Health: Washington, DC, 2014.
9. ISO. *Occupational risk management applied to engineered nanomaterials – Part 2: Use of the control banding approach.* ISO TS 12901-2:2014.
10. OECD. *Harmonized Tiered Approach to Measure and Assess the Potential Exposure to Airborne Emissions of Engineered Nano-Objects and their Agglomerates and Aggregates at Workplaces*. ENV/JM/MONO(2015)19.
11. Marchant, G.; Sylvester, D.; Abbott, K. A New Soft Law Approach to Nanotechnology Oversight: A Voluntary Product Certification Scheme. *UCLA J. Environ. Law Policy* **2010**, *28*, 123–152.

12. Murashov, V.; Howard, J. Protecting Nanotechnology Workers While Waiting for Godot. *J. Occup. Environ. Hyg.* **2013**, *10*, D111–D115.
13. *Nanotechnology Standards*; Murashov, V., Howard, J., Eds.; Springer: New York, 2011.
14. ISO. *ISO/TC 229 Nanotechnologies*; available at http://www.iso.org/iso/home/standards_development/list_of_iso_technical_committees/iso_technical_committee.htm?commid=381983 (accessed on October 5, 2015).
15. OECD. *Publications in the series on the safety of manufactured nanomaterials*; available at http://www.oecd.org/env/ehs/nanosafety/publicationsintheseriesonthesafetyofmanufacturednanomaterials.htm (accessed on October 5, 2015).
16. Howard, J.; Murashov, V. National nanotechnology partnership to protect workers. *J. Nanopart. Res.* **2009**, *11*, 1673–1683.
17. National Institute for Occupational Safety and Health. *Nanotechnology: Field Studies Effort.Workplace Safety & Health Topics*; available at http://www.cdc.gov/niosh/topics/nanotech/field.html (accessed on October 5, 2015).
18. Colleges of Nanoscale Science and Engineering, SUNY Polytechnic Institute. *SUNY Colleges of Nanoscale Science and Engineering and National Institute for Occupational Safety and Health announce expanded partnership*; News Release, October 2, 2014; available at http://www.sunycnse.com/Newsroom/NewsReleases/Details/14-10-02/SUNY_Colleges_of_Nanoscale_Science_and_Engineering_and_National_Institute_for_Occupational_Safety_and_Health_Announce_Expanded_Partnership.aspx (accessed on October 5, 2015).
19. National Institute for Occupational Safety and Health. *General safe practices for working with engineered nanomaterials in research laboratories*; DHHS (NIOSH) Publication No. 2012–147; U.S. Department of Health and Human Services, Centers for Disease Control and Prevention, National Institute for Occupational Safety and Health: Washington, DC, 2014.
20. *NanoHUB. Welcome to the GoodNanoGuide!* Available at https://nanohub.org/groups/gng (accessed on October 5, 2015).

Chapter 13

What Is Responsible Development of Nanotechnology?

Albert Lin*

University of California Davis School of Law, 400 Mrak Hall Drive, Davis, California 95616, United States
***E-mail: aclin@ucdavis.edu.**

In discussions regarding the ethical obligations of nanotechnology researchers, one frequently encounters the phrase "responsible development." But what exactly does responsible development mean? The phrase can refer to conduct that follows ethical rules such as avoiding conflicts of interests and treating research subjects with respect. Alternatively, the phrase may describe a duty among scientists to assess and account for a technology's positive and negative effects, especially its health, safety, and environmental consequences. A third, broader notion of responsible development incorporates public engagement into research and calls on scientists not only to reflect expansively on the ramifications of their work but also to rethink their work in light of public viewpoints and concerns. These multiple understandings of responsible development are appealing in the abstract, but the last in particular raises challenges to implement in practical and concrete ways.

Introduction

In my work as a legal academic, I have wrestled with the question of how society can develop emerging technologies—such as nanotechnology—in ways that advance society's interests while avoiding or mitigating the hazards that new technologies can bring. I have frequently encountered in the literature on nanotechnology innovation the concept of "responsible development of nanotechnology." This concept is the subject of this article: What exactly does "responsible development" mean? And equally important, how can scientists implement responsible development in their everyday work?

All of this presupposes that we should strive for "responsible development of nanotechnology," so let me say a few words on this first. First, responsible development is important as a personal matter, in terms of doing the work that we do for society while attending to health, safety, and environmental concerns. Responsible development is also important for the field of nanotechnology as a whole. With its potential to reshape what we do and how we do it—as well as its potential for harm— nanotechnology must be developed responsibly if its promise is to be fulfilled.

So what is responsible development, and who should be responsible for responsible development? In this article, I sketch out 3 conceptions of responsible development:

- First, responsible development as the ethical conduct of research;
- Second, responsible development as risk assessment and management;
- Third, responsible development as engaging citizens and stakeholders and incorporating their interests into the research & development process.

Although businesses, research institutions, governments, civil society organizations, and the public all have important roles to play in responsible development, scientists are essential to implementing each of these conceptions.

Responsible Development as the Ethical Conduct of Research

Surely, responsible development includes the ethical conduct of research. Research institutions and funding organizations demand this. To take an example, my home institution, UC Davis, has a self-titled "comprehensive research ethics program" that covers topics such as avoiding conflicts of interest, treating research subjects with respect, and properly reporting data and attributing authorship. The ethical conduct of nanotechnology research encompasses these sorts of measures as well as steps to promote safety in the lab, such as taking suitable precautions and avoiding prohibited shortcuts (*1*). The conduct of research safely, with integrity, and free from improper influence is critical to the production of unbiased truth. This first understanding of responsible development is consistent with traditional views of the relationship between scientists and society at large: a relationship characterized by a division of moral labor in which the scientific community enjoys the freedom and resources to pursue basic research, while society decides what to do with the knowledge scientists generate (*2*).

Note that this first understanding of responsible development is a fairly minimalist one. It stresses the ethical conduct of research, but gives little independent meaning to the notion of responsible research. Likewise, it fails to say much about the development aspect of responsible development. If we focus too narrowly on the ethical conduct of research, we may miss the broader social context in which research is occurring and fail to consider how nanotechnology research might differ from other areas of scientific inquiry.

Responsible Development as Risk Assessment and Management

Whether nanotechnology differs from non-nanotechnology science is a complex issue that need not be resolved here. Suffice it to say that nanotechnology's potential for adverse health and environmental consequences demands the reasonable assessment and management of these consequences as part of responsible development (*3*). This second conception of responsible development—responsible development as risk assessment and management—largely tracks the National Research Council's characterization of "responsible development of nanotechnology" as "the balancing of efforts to maximize the technology's positive contributions and minimize its negative consequences." (*4*). Specifically, responsible development "implies a commitment to develop and use technology to help meet the most pressing human and societal needs, while making every reasonable effort to anticipate and mitigate adverse implications or unintended consequences."

Tort law has a role, albeit modest, in implementing this second notion of responsible development. The threat of legal liability deters negligent conduct—i.e., where an actor knew or should have known of unreasonable risk. In addition, strict liability for defective products provides some further protection for product users. In the case of nanomaterials, however, the actual threat of liability is rather limited. This is so because victims face various difficulties both in identifying harm and in causally tracing that harm to a nanomaterial and to the parties ultimately responsible for it.

Responsible development calls us to go beyond the mere avoidance of tort liability or the mere satisfaction of legal responsibilities. It involves a moral responsibility to study and minimize risks in light of nanotechnology's unique characteristics.

One way of describing this understanding of responsible development is in terms of "not blowing things up (*5*)." This is an uncontroversial notion—or at least it should be. Of course we shouldn't blow things up! But figuring out how to operationalize this principle . . . that is subject to disagreement. How much uncertainty are we willing to tolerate regarding the possibility of blowing things up? Should we take a precautionary approach, demanding that specified safety thresholds be met before commercializing a new nanomaterial? Or should we assume that such materials are "innocent until proven guilty?" Is some amount of "things blowing up" tolerable? And if so, how much?

The challenges of assessing nanomaterial risk complicate these issues. Life cycle assessment is the approach generally recommended for evaluating the environmental impacts of products and processes. Life cycle assessment may not always work so well, however, in the case of nanotechnology. Life cycle assessment requires fairly comprehensive information about risks and environmental costs. For nanotechnology, such information is often incomplete. And it may remain so for years (*6*), especially if specific nanomaterials require individualized toxicity characterizations (*7*). Furthermore, even if we had more and better health and safety data, we should not expect life cycle assessments to generate "clean," scientific answers. Life cycle assessments possess important normative dimensions that call for the incorporation of societal values: for example, there are judgment calls to be made about what activities to analyze, which environmental impacts to consider, and what risks matter (*6*).

Ultimately, questions regarding how to proceed with nanotechnology are questions of risk management that cannot and should not be answered by scientists alone. Typically, we look to policymakers to weigh scientific input as well as public concerns, political considerations, costs, and distributional factors as they design laws and regulatory regimes to manage risk (*8*). Policymakers should engage the concerns that nanotechnology raises and respond in a manner reflective of the public interest. At the same time, we must recognize that risk management does not consist solely of legal mechanisms. It also includes voluntary initiatives to assess risks and reduce exposure to untested materials. Even in the absence of legal mandates—and perhaps especially in their absence—scientists have a responsibility to do more.

Efforts like the DuPont/Environmental Defense Nano Risk Framework are valuable and provide an important starting point. The Nano Risk Framework encourages companies and research institutions that are working with nanomaterials and their applications to develop information on their properties, hazards, and exposures. Using this information, as well as appropriate assumptions to fill informational gaps, companies and institutions can proactively manage risks. In the occupational safety and health context, Vladimir Murashov and Chuck Geraci, two of the speakers at this symposium, co-authored a paper describing five criteria to demonstrate responsible development of nanotechnology and elaborating on concrete steps business enterprises can take in regards to each criterion (*9*). These five criteria include: "anticipating, identifying, and tracking potentially hazardous nanomaterials in the workplace; assessing worker's exposures; assessing and communicating hazards; managing risks; and fostering the safe development of nanotechnology."

Notwithstanding the development of such criteria and tools like the Nano Risk Framework, there remains a persistent danger of passing the buck with respect to nanotechnology risks. Unfortunately, one cannot assume that regulators or someone else will analyze the risks. Regulators simply cannot keep up. And in some instances, there will be no one else to analyze the risks. As Lynn Bergeson discussed earlier, informational and regulatory gaps exist and will arise, and the conduct of individual researchers serves as a critical safety net to protect against these gaps. Scientists have an important role to play in an ongoing evaluation of risks and benefits, taking into account the interests of all stakeholders, before and after projects are undertaken (*10*). What is appropriate in each instance will vary, but specific steps that individual researchers may take include: increased vigilance for adverse effects, disclosure of the results of safety tests, minimization of nanomaterial exposure and release, and transparency in the use of nanomaterials.

Responsible Development as Engaging Citizens and Stakeholders and Incorporating Their Interests into the Research and Development Process

Undoubtedly, it is a good idea not to blow things up, and each of the steps just mentioned can help achieve this goal. But surely we aspire to more than simply not blowing things up. This brings us to the third conception of responsible development, a broader conception in which society and scientists are partners in an interactive process of innovation that takes into account more than a technology's direct costs and benefits. There are two fundamental components to this approach. The first, which we might call mindfulness or reflexivity, refers to the undertaking of broader, ethical reflection regarding societal context and implications (*11*). The second component, engagement, would involve scientists engaging in a dialogue with citizens and stakeholders (*11*). These two components—mindfulness and engagement—should occur as integral parts of the research process. If done afterwards, it may be too late: a technology may already be entrenched, or irreversible effects may have already occurred (*12*). A more likely scenario is that these components will not be addressed at all. The temptation for individuals is to assume that someone else is responsible for worrying about technology's broader implications.

This third kind of responsible development is harder. It imposes moves beyond a negative obligation to avoid harm. It's amorphous: risk assessment is something we often can quantify, and readily can wrap our arms around. Mindfulness and engagement are not so easily encapsulated—they sound like things you practice in a commune or on a retreat while singing "Kumbayah," not things you do in a lab. Furthermore, mindfulness and engagement lie beyond scientists' comfort zone: they require transparency and engagement with non-scientists regarding what is being done in the lab and why.

The rest of this article suggests some concrete ways to carry out this third conception of responsible development. Note that these ideas are not original to me. Nor have these ideas been subject to rigorous scientific testing. Rather, these ideas reflect various things that are being tried and that may be worth integrating into your work.

Responsible development is one aspect of a broader concept: corporate social responsibility, or CSR. It's worth spending a moment to discuss CSR because its proponents and practitioners have given some attention to the question of how to implement it. The underlying premise of CSR is that corporations receive sanction from society to operate, and in exchange for that sanction, they have a corresponding responsibility to society to contribute to its betterment. Though CSR lacks a standard definition, the basic idea is that companies should not just be selling goods and services and complying with applicable laws: they also should voluntarily account for the impacts of their activities on society and the environment through transparent and ethical behavior (*13*). In implementing CSR, companies do face various challenges, including how to communicate values within the company and beyond to its suppliers, and how to incorporate these values into concrete practices (*14*). Measures to implement CSR may involve, for example, establishing metrics for performance based on environmental factors such as recycled content or energy use, setting goals with respect to these factors, and creating incentives for meeting these goals.

CSR is typically applied by firms with known products and markets (*14*). But it also can be incorporated into emerging technologies such as nanotechnology. To address the challenges of implementing responsible development in this context, companies can turn to tools such as the Nano Risk Framework (*15*). They can also look to codes of conduct, which provide guidance to individual researchers as well as to institutions. Admittedly, the principles contained in these codes are often general and not intended to serve as "an auditable set of standards."(*16*). However, the codes often also include concrete suggestions that one can adopt.

For example, the European Commission's Code of Conduct for Responsible Nanosciences and Nanotechnologies Research counsels researchers to take a cautionary approach in setting research priorities. One specific recommendation is for researchers to avoid "research involving deliberate intrusion of nano-objects into the human body, or their inclusion in food, feed, toys, cosmetics and other products that may lead to exposure to humans and the environment" "as long as risk assessment studies on long-term safety is not available."(*17*).

Another code of conduct, Responsible NanoCode, sets out seven overarching principles. Among the concerns these principles address are stakeholder involvement; worker safety; and social, environmental, health, and ethical implications (*16*). The code then elaborates on these principles through numerous examples. To promote the safety of workers and the public, for instance, organizations may publicly disclose information regarding when nanomaterials have been used, to what extent their safety has been evaluated, and how they might be safely handled (*16*). Similarly, to protect public health and the environment, the Responsible NanoCode recommends that organizations put in place processes to evaluate risks and support government and independent initiatives to bridge informational gaps (*16*).

The measures I have mentioned so far would encourage researchers or research institutions to reduce potential risks. Let me now turn to the issue of engaging stakeholders and the public. The general question of public engagement asks how "actors from science and society . . . [might] work together as . . . partners in order to produce better results"? (*18*). In the context of nanotechnology, we might sharpen our inquiry to ask: what concrete steps might be taken towards broad collaboration involving not just chemists and toxicologists to understand the effects of nanomaterials, but also social scientists and the public in evaluating the value of nanotechnology research and applications? (*4*).

The literature on public engagement with emerging technologies describes a wide variety of tools, including focus groups, consensus conferences, citizens' juries, science cafes, and even cultural festivals. (*19*). To figure out which tools may make sense for nanotechnology researchers, we can again turn to the European Commission's Code of Conduct for Responsible Nanosciences and Nanotechnologies Research and the Responsible Nanocode. The European Commission's Code encourages researchers "to consider, at the earliest stages and through participatory foresight exercises, the future implications of technologies or objects being researched." (*17*). "Consultation with relevant ethics committees" is one means the code suggests for implementing this recommendation (*17*).

Moving beyond such consultation, companies and institutions can conduct meetings with employees, customers, civil society organizations, and the general public; participate in programs and panels with various stakeholders; engage and train suppliers, or initiate web forums (*16*). In undertaking any of these activities, companies will have to strike a balance between the desire to promote meaningful input and engagement and valid concerns about protecting confidential business information and intellectual property (*20*).

Another possibility involves the integration of social scientists into the lab and into research teams. Erik Fisher & Daan Schuurbiers have conducted laboratory engagement studies illustrating how this might occur: ethicists or other social scientists participate in a series of conversations and interactions with nanotechnology researchers during routine research and innovation activities (*21*). The social scientists had to possess or acquire sufficient technical understanding to interact intelligently and deeply with the researchers, of course (*22*). Fisher & Schuurbiers found that the resulting interactions not only enhanced critical reflection among lab researchers but also yielded tangible effects on lab practices and research agendas (*21*). Beyond the scope of their study, such efforts suggest the possibility of what science & technology scholars describe as "upstream modulation"—interaction that spurs new, socially responsible lines of research—as well as "midstream modulation"—interaction on ongoing projects that influences such research going forward (*22*).

Some of the efforts just described may require resources that are simply unavailable, or may involve operational or institutional changes that seem infeasible. However, these obstacles need not prevent the practice of the third conception of responsible development in terms of engaging citizens and stakeholders and incorporating their interests into the research & development process. Ultimately, to quote Kamilla Kjolberg and Roger Strand, "[r]esponsible nanoresearch . . . may be not to put away pertinent concerns raised in [the course of one's] own contemplation, in dialogue with peers or by other stakeholders, even if not able to solve them, but to remain committed to thinking through these kinds of questions on a regular basis." (*2*). In the work that we do, each of us can periodically reflect on questions such as: "Do we need that use anyway?" (*3*), "Who might be affected by this work," and "What are its probable or possible long-range consequences?" (*18*). Such reflection can promote responsible development – broadly understood in terms of one's responsibility to society to wisely develop and deploy the powerful tools that nanotechnology may provide.

References

1. McGinn, R. Ethical Responsibilities of Nanotechnology Researchers: A Short Guide. *Nanoethics* **2010**, *4*, 1–12.
2. Kjolberg, K. L.; Strand, R. Conversations About Responsible Nanoresearch. *Nanoethics* **2011**, *5*, 99–113.
3. Shelley-Egan, C.; Davies, S. R. Nano-Industry Operationalizations of 'Responsibility': Charting Diversity in the Enactment of Responsibility. *Rev. Policy Research* **2013**, *30*, 588–604.

4. National Research Council. *A Matter of Size: Triennial Review of the National Nanotechnology Initiative*; National Research Council: Washington, DC, 2006.
5. Davies, S. R.; Glerup, C.; Horst, M. In *Responsibility in Nanotechnology Development*; Arnaldi, S., Ferrari, A., Magaudda, P., Marin, F., Eds.; The International Library of Ethics, Law and Technology; Springer: Dordrecht, 2014; Vol. 13, pp 143−159.
6. Wender, B. A.; Foley, R. W.; Hottle, T. A.; Sadowski, J.; Prado-Lopez, V.; Eisneberg, D. E.; Laurin, L.; Seager, T. P. Anticipatory Life-Cycle Assessment for Responsible Research and Innovation. *J. Responsible Innovation* **2014**, *1*, 200–207.
7. Fleischer T.; Jahnel, J.; Seitz, S. B. Technology Assessment Beyond Toxicology: The Case of Nanomaterials. In *Responsibility in Nanotechnology Development*; Arnaldi, S., Ferrari, A., Magaudda, P., Marin, F., Eds.; The International Library of Ethics, Law and Technology; Springer: Dordrecht, 2014; Vol. 13, pp 79−96.
8. Doremus, H.; Lin, A.; Rosenberg, R. *Environmental Policy Law*, 6th ed.; Foundation Press: New York, 2012.
9. Schulte, P. A.; Geraci, C. L.; Murashov, V.; Kuempel, E. D.; Zumwalde, R. D.; Castranova, V.; Hoover, M. D.; Hodson, L.; Martinez, K. F. Occupational Safety and Health Criteria for Responsible Development of Nanotechnology. *J. Nanopart. Res.* **2014**, *16*, 2153.
10. Colussi, I. A. The Role of Responsible Stewardship in Nanotechnology and Synthetic Biology. In *Responsibility in Nanotechnology Development*; Arnaldi, S., Ferrari, A., Magaudda, P., Marin, F., Eds.; The International Library of Ethics, Law and Technology; Springer: Dordrecht, 2014; Vol. 13, pp 53−72.
11. Grunwald, A. Responsible Research and Innovation: An Emerging Issue in Research Policy Rooted in the Debate on Nanotechnology. In *Responsibility in Nanotechnology Development*; Arnaldi, S., Ferrari, A., Magaudda, P., Marin, F., Eds.; The International Library of Ethics, Law and Technology; Springer: Dordrecht, 2014; Vol. 13, pp 191−205.
12. Collingridge, D. *The Social Control of Technology*; St. Martin's Press: New York, 1980.
13. ISO. *ISO 26000: 2010 Guidance on Social Responsibility*; https://www.iso.org/obp/ui/#iso:std:iso:26000:ed-1:v1:en (accessed on April 30, 2015).
14. Bos, C.; van Lente, H. Value Chain Responsibility in Emerging Technologies. In *Responsibility in Nanotechnology Development*; Arnaldi, S., Ferrari, A., Magaudda, P., Marin, F., Eds.; The International Library of Ethics, Law and Technology; Springer: Dordrecht, 2014; Vol. 13, pp 129−141.
15. Environmental Defense-DuPont Nano Partnership. *Nano Risk Framework*, 2007.
16. Insight Investment, Royal Society, Centre for Process Innovation, Nanotechnology Industries Association. *Responsible Nano Code: Information on the Responsible Nano Code Initiative*, 2008.

17. European Commission. *Code of Conduct for Responsible Nanosciences and Nanotechnologies Research*, 2008.
18. Glerup, C.; Horst, M. Mapping 'Social Responsibility' in Science. *J. Responsible Innovation* **2014**, *1*, 31–50.
19. Delgado, A.; Kjolberg, K. L.; Wickson, F. Public Engagement Coming of Age: From Theory to Practice in STS Encounters with Nanotechnology. *Public Understand. Sci.* **2011**, *20*, 826–845.
20. Kuzma, J.; Kuzhabekova, A. Corporate Social Responsibility for Nanotechnology Oversight. *Med. Health Care Philos.* **2011**, *14*, 407–419.
21. Fisher, E.; Schuurbiers, D. Socio-technical Integration Research: Collaborative Inquiry at the Midstream of Research and Development. In *Early Engagement and New Technologies: Opening Up the Laboratory*; Doorn, N., Schuurbiers, D., van de Poel, I., Gorman, M. E., Eds.; Springer: Dordrecht, 2013; Vol. 16, pp 97–110.
22. Gorman, M. E.; Calleja-López, A.; Conley, S. N.; Mahootian, F. Integrating Ethicists and Social Scientists into Cutting Edge Research and Technological Development. *Early Engagement and New Technologies: Opening Up the Laboratory*; Doorn, N., Schuurbiers, D., van de Poel, I., Gorman, M. E., Eds.; Springer: Dordrecht, 2013; Vol. 16, pp 157–173.

Editors' Biographies

H. N. Cheng

H. N. Cheng (Ph.D., University of Illinois) is currently a research chemist at Southern Regional Research Center of the U.S. Department of Agriculture in New Orleans, where he works on projects involving improved utilization of commodity agricultural materials, green chemistry, and polymer reactions. Prior to 2009 he was with Hercules Incorporated where he was involved (at various times) with new product development, team and project leadership, new business evaluation, and analytical research. Over the years, his research interests have included NMR spectroscopy, polymer characterization, biocatalysis, functional foods, pulp and paper technology, and green polymer chemistry. He is an ACS Fellow and a POLY Fellow and has authored or co-authored 208 papers, 25 patent publications, co-edited 13 books, and organized or co-organized 28 symposia at national ACS meetings since 2003. He is active in the ACS at local, division and national levels. He has served in the following ACS national committees: CEPA, SOCED, ConC, IAC, and NTS, including chairs of CEPA and IAC.

Laurence J. Doemeny

Laurence J. Doemeny (Ph.D., University of California, Santa Barbara) is retired from the U.S. Public Health Service Commissioned Corps. He spent his career at the National Institute for Occupational Safety and Health in various research and management capacities. His research included aerosol physics, techniques to measure worker exposure to chemicals in the workplace, miniaturization of analytical instruments, quality assurance. In management, he was involved with management operation, security, technology transfer, labor relations, building and laboratory design, served on the Greater Cincinnati Federal Executive Board (including chair), Adjunct Professor at the University of Cincinnati. He has authored, co-authored or contributed to numerous peer reviewed papers, government publications and books. In addition to service related awards, he is an AAAS, ACS and DCHAS Fellow. Over the years, he served on the ACS Government Affairs Committee, the Committee on Chemical Safety (chair) and the Ethics Committee.

Charles L. Geraci

Charles L. Geraci, (Ph.D., Michigan State University) is Associate Director for Nanotechnology, National Institute for Occupational Safety and Health (NIOSH). He has 40 years of Industrial Hygiene practice experience, including Director of a regional laboratory for DataChem Laboratories; Associate Director of HS&E at the Procter & Gamble Company; and Director of Industrial Hygiene for two consulting firms. He is certified by the American Board of Industrial Hygiene in both the Comprehensive Practice and the Chemical Aspects of Industrial Hygiene and is a Fellow of the American Industrial Hygiene Association. He provides overall coordination and strategic guidance to the nanotechnology research program at NIOSH and collaborates internationally with other country programs on various aspects of nanotechnology workplace safety and health. He directly manages a number of NIOSH nanotechnology projects that focus on the development and dissemination of workplace risk management guidelines, including the document "Approaches to Safe Nanotechnology." Dr. Geraci sponsors the NIOSH nanotechnology field team that is conducting visits to nanomaterial producers and users to characterize exposures, evaluate controls, and develop best practices.

Diane Grob Schmidt

Diane Grob Schmidt (Ph.D., University of Cincinnati), the 2015 ACS President, was an Executive at The Procter & Gamble Company, where she served as a R&D Section Head for 17 years. Her P&G career spanned 1981-2014 during which she played key roles in such brands as Tide®, Head & Shoulders®, Pert Plus® and Safeguard®. Dr. Schmidt had responsibility for health, safety and environment, and regulatory affairs before retiring from P&G in 2014. She is currently an Adjunct Professor in the Department of Chemistry at the University of Cincinnati. She is the inventor or co-inventor on myriad patents, in addition to author or co-author on chemistry publications in refereed journals. She has received many awards, including ACS Fellow, Fellow of the Division of Chemical Health & Safety, Henry Hill Award, and Distinguished Scientist of Cincinnati (first woman so honored). She has served on the editorial boards of the Journal of the Society of Cosmetic Chemists and the Journal of Chemical Health & Safety. She has been an ACS member since 1968 and held a wide variety of leadership positions, including three consecutive terms on the Board of Directors. As 2015 ACS President, her Presidential theme was "Inspiring and Innovating for Tomorrow". Her legacy as ACS President includes: U.S. and Global Grand Challenges via impactful programming (Nanotechnology, Energy, BRAIN Initiative/Chemical Measurements/Chemical Neurotransmission), championing the establishment of the American Association of Chemistry Teachers (AACT), advocacy on behalf of ACS members and focus on industry and industrial ACS members.

Indexes

Author Index

Subject Index

O

R

S

Printed in the USA/Agawam, MA
December 5, 2017

664506.010